Breaking Down the Barriers: Electrokinetics for Bioremediation

Alisha

Table of content

1. Scope and Outline of the book

As a central quantitative indicator of biodegradation of chemicals, bioavailability denotes the 'degree of interaction of chemicals with living organisms'. For contaminant bioremediation, bioavailability is determined by several processes: the 'availability to a degrader' (the release and transport of contaminants to degrader), and the 'activity of a degrader' (cell's uptake and rate of biodegradation). The initial step of improving bioavailability is therefore to ensure the availability of contaminants. Two strategies may give solutions to low bioavailability resulting from limited contaminant availability to degrader: i) enhancing bacterial deposition and transport to contaminants; ii) enhancing contaminant release from sources and transport to contaminant degraders. There is hence interest in enhancing bioavailability by affecting the contaminant availability following these two strategies. One powerful tool effective in both strategies is the electrokinetic technique. The main objective of the book was to investigate the principles of electrokinetic effects on the bacterial deposition and contaminant release to improve bioavailability. The drivers of bioavailability and the main questions of this book are explained in **Chapter 1.**

Chapter 2 gives a general introduction to the concept of electrokinetic approaches to overcome the bottleneck of bioavailability. One of the most important initial steps of improving bioavailability is enhancing the availability of contaminants to microbes. Electrokinetic approaches are therefore powerful to improve the bioavailability by two strategies: i) controlling bacterial deposition and transport; ii) controlling contaminant sorption and desorption.

Chapter 3 describes the investigations of electrokinetic effects on bacterial transport in porous media. Applying percolated laboratory columns, we observed the deposition efficiency of four bacterial strains in porous media with the presence and absence of electric fields. The driver of electrokinetic effects was further investigated by correlating the variations of deposition efficiency to forces acting on bacteria. An approach to estimate the electrokinetic effects on bacterial deposition was established to predict the effects of electric fields on bacteria strains with different physicochemical properties.

Chapter 4 describes the investigations of electrokinetic effects on bacterial deposition on planar surfaces. Applying the quartz crystal microbalance with dissipation monitoring (QCM-D) system together with microscope observations after deposition, the deposition rate and rigidity in a wide electrolyte concentration range were real-time observed in the presence and absence of DC electric fields. The approach to estimate electrokinetic effects on bacterial deposition was evidenced by correlating deposition rates observed by microscope counting, the deposition rigidity measured by QCM-D, and theoretically calculated net forces.

Chapter 5 describes the investigations of electrokinetic effects on contaminant sorption and desorption on sorbents. Applying a column reactor filled with sorbents of different physiochemical properties, the sorption and desorption rates of phenanthrene were observed in the presence and absence of electric fields. An approach to estimate electrokinetic effects on contaminant release was established by correlating the sorption/desorption rate to the relative strength of electroosmotic flow and the sorption Gibbs free energy.

Chapter 6 summarizes the main findings of the book and relates them to the major objective of this book and their relevance for environmental applications. Possibilities of estimating the improvement of contaminant availability applying the two established electrokinetic-effect approaches in natural and manmade systems were discussed, the research interests that can be conducted following current results were also proposed.

2. General Introduction

2.1 Driving Factors of Contaminant Bioavailability

As a central indicator of the biodegradation of chemicals, the term 'bioavailability' is usually used to quantitatively describe the 'degree of interaction of chemicals with living organisms'[1]. In previous research, several biological and chemical methods for assessing bioavailability have been described[2]. For contaminant biodegradation, Bosma et al.[3] defined bioavailability as the rate of a contaminant's mass transfer to microbial cells relative to their intrinsic catabolic potential to degrade the contaminant. This perspective points at the relevance of mass fluxes for 'degradation processes' and discriminates bioavailability for degradation from bioavailability for 'non-degradation' processes that lead to poisoning or inhibition of the receptor organisms.

Several processes (Fig. 1) determine the bioavailability of contaminants: the release and transport of the contaminant from the source, the mobility of the degrader cell to the contaminant (both of these processes determine the 'availability to degrader'), and the cell's uptake and rate of biodegradation ('activity of degrader'). Low bioavailability for biodegradation may arise if the release rate of contaminant from the source is low and the mobility of bacteria is limited[2]. Therefore, one initial step to derive productive biodegradation (i.e., high bioavailability) is to ensure the availability of contaminant to microbes.

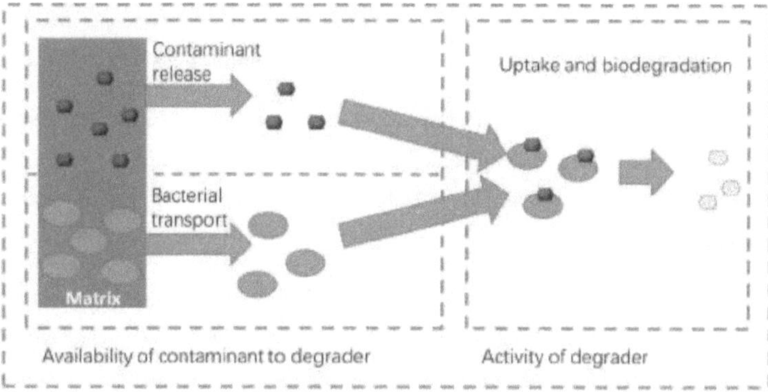

Figure 1. Conceptualization of the main processes driving the bioavailability of contaminant. Bioavailability is controlled by the availability of contaminant to degrader and the activity of degrader. The availability of contaminant is determined by the contaminant release and the bacterial transport in the matrix.

Two different strategies are adopted to overcome the bottleneck of bioavailability caused by the constraints of contaminant availability. Firstly, developing methods inducing effects on bacterial deposition or transport to improve contaminant availability. For instance, in extreme groundwater habitats[4] or disturbed soil ecosystems with very low bacterial concentration[5,6], dispersing contaminant degraders into the system is needed and increasing the deposition rate will enhance contaminant availability. Inversely, in low permeability soil systems, increasing the transport of degrader cells will increase the possibility of microbes to reach contaminants adsorbed on soil pores therefore may enhance contaminant availability[7,8].

Secondly, developing methods inducing effects on contaminant sorption or release to improve contaminant availability. For instance, in systems containing enough biomass, the release of contaminant will enhance the availability of contaminants[9]. Inversely, low contaminant concentration will limit the habitation of degraders when it cannot meet the maintenance requirements of the degrader population[10]. Adsorption of contaminants on the matrix will then enhance the availability of contaminants to biofilm habitats on the matrix.

2.2 Electrokinetic Improvement of Contaminant Bioavailability

According to the two strategies to improve bioavailability, several approaches such as electrokinetics[11–13], surfactants[14–16], pressure-driven hydraulic flow flushing[17–20] have been developed to affect the bacterial deposition and contaminant release. Electrokinetic approaches showed their advantages in controlling bacterial deposition and contaminant release at the same time[21–26], with no harmful effects on the viability of degrading bacterial cells[27].

2.2.1 Electrokinetic Effects on the Microbe-Matrix-Contaminant Interactions

Electrokinetic bioremediation is based on the application of an electric field to the contaminated ecosystem by a series of electrodes designed as anodes and cathodes, which induce a variety of electrokinetic phenomena to enhance the mobilization and transport of the contaminants for subsequent contaminants removal[22].

In contaminated ecosystems, there are mutual interactions between matrix, contaminants, and microorganisms (cf. the base triangle in Fig. 2). These interactions form the base triangle in Fig. 2 that includes: i) microorganism deposition and transport in the matrix, ii) contaminant sorption/desorption interactions with the matrix, iii) contaminant degradation by microorganisms. Electrokinetic phenomena including electroosmosis and electrophoresis are powerful tools in controlling the movement of bacteria and (bio-)colloidal particles with an external direct current (DC) electric field[28–31].

The application of the electric field therefore may stimulate the base triangle interactions, which can be described with a tetrahedron by performing electric field effects on the base triangle mutual interactions (Fig. 2). The electric field effects include three aspects: i) electric field effects on bacterial deposition and transport; ii) electric field effects on the interaction of contaminants with matrix; iii) electric field effects on the biodegradation efficiency. In this study, we concentrate on the electrokinetic effects on bacterial deposition and contaminant release which are essential to overcome the constraining factors of availability, as the first step to improve bioavailability.

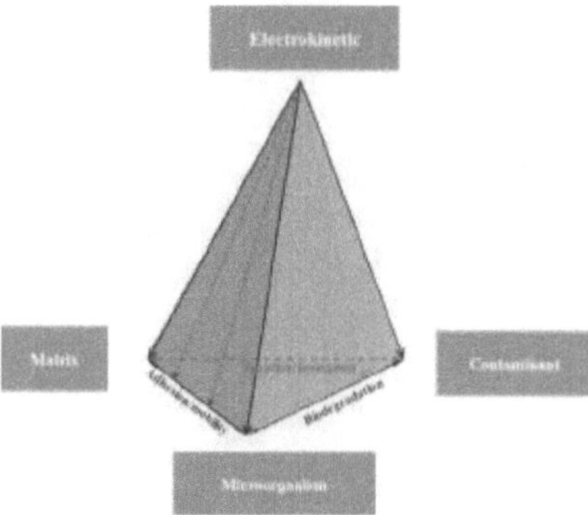

Figure 2. Schematic of electrokinetic effects on the interactions in the soil system (adapted from Wick et al.[11])

2.2.2 Principles and Applications of Electrokinetic Phenomena

The main power of the electric field on the mobilization of particle fluids and molecules comes from electrokinetic phenomena. In direct current (DC) electric fields, there are distinct electrokinetic phenomena whenever there is relative motion between the liquid phase and the solid phase: electromigration, electroosmosis, electrophoresis, streaming potential, and sedimentation potential[32], the dominating electrokinetic phenomena depend on the way in which relative motion is induced. When an external DC electric field is applied to the electrolyte, the electrical force causes motion of ions and charged particles, resulting in mainly two electrokinetic phenomena electroosmosis and electrophoresis, which play key roles in the interactions at the liquid-solid interface. To investigate the effects of these two electrokinetic phenomena, it is important to first understand their origin, that is, the surface charge and electrical potential distribution adjacent to the solid surface (i.e., the electrical double layer).

When a solid phase material is placed in a polar liquid, dipolar molecules in the aqueous phase will tend to be oriented in a particular direction at the interface and generate a potential difference[33]. The surface charge must be exactly balanced by an equal and opposite charge in the solution. Therefore, the balancing charge is accounted for by an excess number of oppositely charged ions in the solution adjacent to the solid surface and a deficit of similarly charged ions or co-ions. This will produce three layers according to the ion distribution along with the separation distance: Stern layer, diffuse layer, and bulk liquid, with the first two dense layers being defined as the electrical double layer (Fig. 3). The potential from the solid surface to the distance of the diffuse layer is defined as zeta potential[32–34], it has been widely used to characterize the charge property due to plenty measuring methods[35–37].

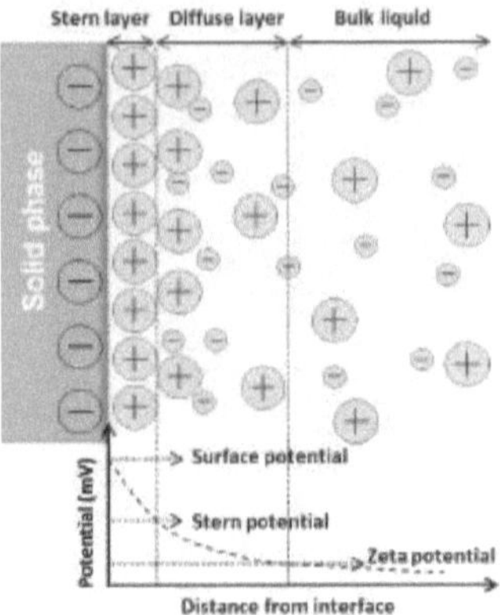

Figure 3. Schematic of the electrical double layer and potential distribution (adapted from Elimelech et al.[32])

Potential distribution over the increment of distance from the solid-liquid interface is highly related to the thickness of the electrical double layer (κ^{-1}), which can be calculated by equation[38]

$$\kappa^{-1} = (\sqrt{\frac{2000F^2}{\varepsilon_0 \varepsilon_r RT}} \times I)^{-1} \tag{1}$$

where F is the Faraday charge (i.e. the magnitude of the charge on a mole of electrons = 96485 coulombs), ε_0 is the permittivity of free space, ε_r is the dimensionless dielectric constant or relative permittivity, R is the gas constant, T is temperature, I is the ionic strength[32]. The electrical double layer thickness κ^{-1} therefore decreases with the increment of electrolyte ionic strength. The concepts of the electrical double layer and zeta potential can further help to interpret the electrokinetic phenomena, which are the relative movements of ions and particles in the presence of an external DC field.

In the presence of a DC external electric field, the relative motion of counter-ions attached on the solid surface towards the oppositely charged electrode results in the movement of the liquid adjacent to the solid surface, the liquid flow is electroosmotic flow (EOF). EOF is parallel to the solid surface towards the electrode that has the same charge as the solid surface, it rises from zero to a maximum value $V_{EOF,\ max}$ at a short distance (in the range of electrical double layer thickness) above the solid surface, the velocity profile has been investigated in capillary and porous media in previous research[39–43]. $V_{EOF,\ max}$ can be calculated by the equation

$$V_{EOF,max} = -\frac{\varepsilon_0 \varepsilon_r \zeta_C \times E}{\eta} \tag{2}$$

where ζ_C is the zeta potential of solid surface, E is the electric field strength (V m^{-1}), η is the viscosity of the liquid. The profile of electroosmotic flow velocity regarding the increment of separating distance can be described by the following equation with the combination of a simplified EOF expression of the Navier-Stokes equation with the potential distribution described by the Gouy-Chapman model:

$$V_{EOF} = -V_{EOF,max}(1 - \frac{2I_1(\kappa h)}{\kappa r I_0(\kappa h)}) \tag{3}$$

I_0 and I_1 are the zero-order and first-order modified Bessel functions, h is the distance from the solid surface, r is the radius of the channel. The electrical double layer affects the electroosmotic flow velocity, the thinner double layer thickness (higher electrolyte ionic strength), in the shorter distance electroosmotic flow can reach the highest velocity.

Electroosmosis approaches[24,26,44] have shown high potential to mobilize slow-releasing hydrophobic organic pollutants to desorb from matrices of low permeability[45–51]. When an electric field is applied to a matrix immersed in an ionic solution, it invokes an electroosmotic transport process, that is, the surface charge-induced movement of pore fluids in an electric field is usually directed from the anode toward the cathode[52]. It originates from the enrichment of ions in the so-called electric double layer near a surface and is particularly effective in fine-grained materials where macro- and micro-pores dominate. These are situations where the hydraulic flow is extremely small and molecular diffusion may limit the access of sorbates to and the release of sorbates from smaller pores[12,53]. Electroosmotic perfusion induces efficient liquid flow in inter- and intra-particle network pore channels and, hence, increase release rates and natural attenuation of PAH at locations where normal pressure-driven pump and treatment approaches may be inadequate[53–55] or energetically ineffective[56]. Electroosmotic flow (EOF) can thus be applied for the dispersal and separation of uncharged entities or the dewatering of matrices. Contrary to the parabolic velocity profile of pressure-driven hydraulic flow in a pore, the velocity profile of EOF is quasi planar beginning at the electrical double layer located at a few nanometers above the surface. It thus likely arises at scales relevant to contaminant-sorbent interactions. This effect is, for instance, used in capillary electrochromatography (CEC) where EOF (rather than pressure-driven-flow such as in HPLC) is used to effectively separate uncharged solutes between a mobile and a stationary phase[57]. Hassan et al. found electroosmotic flow can be adopted to stimulate desorption of phenanthrene (PHE) from kaolin due to the strong shear produced by electroosmotic flow[56]. On the other hand, recent work showed that DC fields increased PHE sorption rates in carbonaceous exfoliated graphite sevenfold and reduced the PHE desorption rate by > 99%. This was discussed as a result of electroosmotic perfusion of PHE to pores that contribute most of the sorption sites, but are difficult to access in the absence of EOF by molecular diffusion only. Therefore, there are high interests to investigate the mechanism of how electroosmotic flow affects phenanthrene sorption onto and desorption from different types of matrices.

Simultaneously, the movement of charged colloidal particles and molecules (e.g., bacterial cells) to the electrode of the opposite charge in the DC electric field is electrophoresis. Electrophoretic velocity V_{EP} can be quantified by the Smoluchowski equation and Henry's function:

$$V_{EP} = \frac{2\varepsilon_0\varepsilon_r\zeta_p E}{3\eta} f(\kappa a) \tag{4}$$

Where ζ_p is the zeta potential of the target particle, $f(\kappa a)$ is a factor of the electrical double layer parameter κ and channel radius a. Previous research found that $f(\kappa a)$ levels to 1.5 for big particles and high ionic strength while levels to 1.0 for very small particles in low ionic strength electrolyte solutions[32]. Zeta potential can be derived based on this equation with electrophoretic mobility which can be determined by applying dynamic light scattering technique[36,58].

In natural ecological systems, matrix and bacteria strains are generally negatively charged due to the ion-affinity difference of the two phases or the ionization of surface chemical groups. Especially, both Gram-positive and Gram-negative bacteria carry a negative surface charge. For Gram-positive bacteria, the negative charge originates from ionization of phosphate in teichoic acids that linked to either the peptidoglycan or to the underlying plasma membrane; for Gram-negative bacteria, the outer covering of phospholipids and lipopolysaccharides import a strong negative charge.

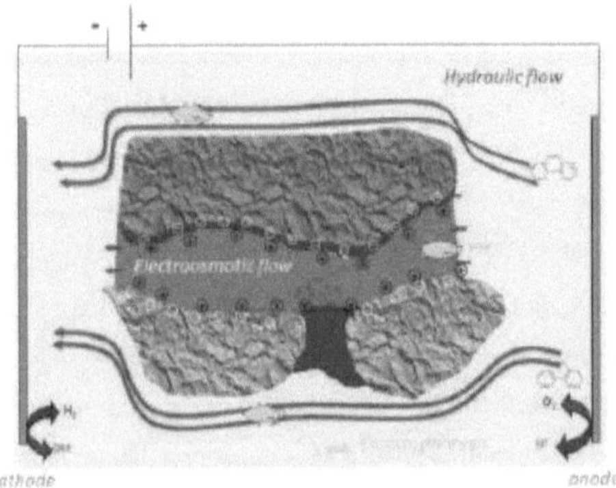

Figure 4. Schematic of electroosmotic flow, electrophoresis, and hydraulic flow

As described above, the electroosmotic flow thus directs towards the cathode while electrophoresis directs towards the anode when both bacteria and solid matrix carry negative charges (Fig. 4). In this case, two powerful electrokinetic phenomena will contradict each other, the dominating phenomenon therefore depends on the relative strength of electroosmosis and electrophoresis. Based on the principles on how electroosmosis and electrophoresis play their roles, investigations were further designed to study the principles of 'electrokinetic effects on bacterial deposition and transport' in the following chapter 2.3 and 'electrokinetic effects on contaminant sorption and desorption' in chapter 2.4, separately.

2.3 Electrokinetic Effects on Bacterial Deposition and Transport

Bacterial deposition and transport are fundamental processes in microbial ecology and biotechnology[59], which enable microbial functions in disturbed systems[8] or promote the formation of biofilms as a major life form of bacteria. While the catabolic activity of biofilms provides essential ecosystem services in natural and man-made systems (e.g. for the degradation of anthropogenic chemicals or in wastewater treatment). There is, hence, strong interest in measures to control microbial deposition to surfaces as the first step in the formation of biofilms.

Bacterial deposition efficiency can be quantitatively investigated using packed percolation columns or by QCM-D monitoring coupled to microscopy cell counting, which in the meanwhile allows for characterizing the rigidity of deposition. In a porous system, deposition to collector surfaces can be investigated by packed column systems (cf. Fig. 5) and quantitatively approximated by clean-bed filtration theory[60,61]. The clean-bed filtration model has been used for calculating bacterial deposition efficiency (i.e., the rate of bacterial retention on a single collector) with overall consideration of Brownian diffusion, interception, and sedimentation.

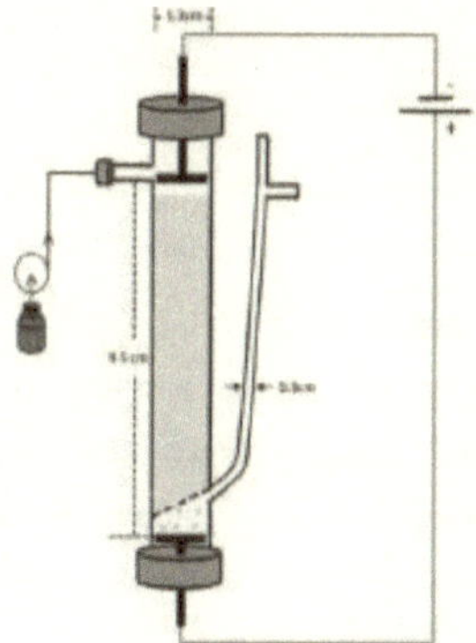

Figure 5. Schematic of the column reactor (adapted from Qin et al.[26])

Furthermore, based on the results in porous media, electrokinetic effects on the deposition of two typical bacterial strains *P. putida* KT2440 and *P. fluorescens* LP6a were further studied with a high accuracy quartz crystal microbalance with dissipation monitoring (QCM-D) system combined with microscope counting in various electrolyte concentrations (cf. Fig. 6). The microscopic technique can be used to evaluate bacterial deposition efficiency, frequency shift, dissipation shift, and deposition rigidity signals can be used to evaluate the mass and rigidity of bacterial deposition. In both porous media and QCM-D system, by comparing the bacterial deposition efficiency with the presence and absence of electric fields, the variations of deposition efficiency and variations of external electrokinetic factors were interlinked and analyzed to investigate the principles of electrokinetic effects.

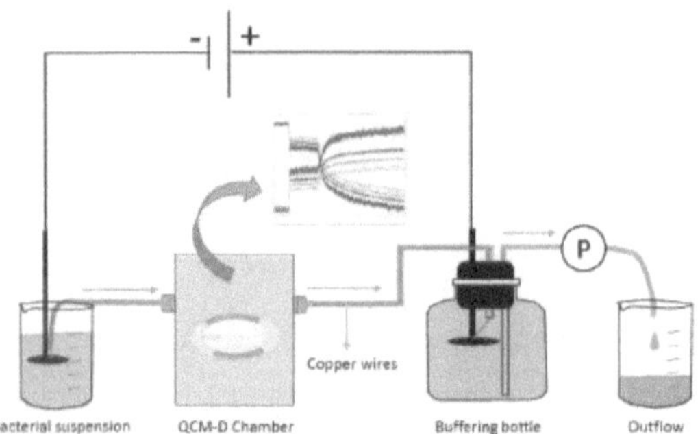

Figure 6. Schematic of QCM-D experimental set-up used in this book

In both systems, bacterial deposition efficiency is influenced by hydraulic flow velocity, physicochemical properties of the microbe, the collector surface, and the aqueous medium[62]. These influencing factors can be quantitatively included in the forces acting on the bacterial cells. In the absence of electric field, bacterial deposition rate is dominated by the electrostatic repulsive energy and Lifshitz-van der Waals energy, which can be quantified by DLVO theory (deduced by Derjaguin, Landau, Verwey, and Overbeek) regarding the separating distance between bacterium and collector surface[32]. With the presence of an external DC field on an ionic solution in a solid matrix, it invokes electrophoresis and electroosmosis, which induce electrophoretic drag force and electroosmotic shear force in the opposite directions as described above[63]. Due to the plug shape flow profile of EOF and drag force of electrophoresis, they have been found to be efficient at the 2nd minimum distance where bacterial deposition interaction takes place and thus significantly affects bacterial deposition efficiency[26]. Inspired by such observations, the dominating forces including DLVO force (F_{DLVO}), hydraulic shear force (F_{HF}), electroosmotic flow shear force (F_{EOF}), and electrophoretic force (F_{EP}) can be further quantified and summed up to a net force. The net force quantitatively includes the effects of bacterial physiochemical properties and the electrokinetic factors and thus can be interlinked to the quantified bacterial deposition efficiency (from both porous media and QCM-D system) to identify the drivers of DC-effects.

2.3.1 Electrokinetic Effects on Bacterial Transport Through Porous Media

The deposition and transport of bacterial cells in porous media was quantified by the clean-bed filtration model, which quantitatively describes the deposition rate by deposition efficiency α_t[26,27], defined as the ratio of the rate of attachment (η_t) to the rate of bacterial transport to the surfaces (η_{trans})[61]:

$$\alpha_t = \frac{\eta_t}{\eta_{trans}} \tag{5}$$

Values of η_t can be calculated from C/C$_0$ values obtained in column experiments

$$C = C_0 \exp\left(-\frac{3(1-p)}{4a_s}\eta_{trans}\alpha_t L\right) \tag{6}$$

where C is the effluent cell concentration, C_0 the influent cell concentration, p the porosity of the packed bed, a_s the radius of the glass beads, L the length of the column, and η_t the transport of bacteria from the solution to the glass surface in the whole experimental time. η_{trans} was approximated by applying the solution to the convection-diffusion equation

$$\eta_{trans} = A_s\left(\frac{A_{132}}{9\pi\eta a_b^2\mu}\right)^{0.125}\left(\frac{a_b}{a_s}\right)^{1.875} + 0.00338A_s\left(\frac{2a_b^2(\rho_b-\rho_l)g}{9\eta\mu}\right)^{1.2}\left(\frac{a_b}{a_s}\right)^{-0.4} + 4A_s^{0.33}\left(\frac{12\mu a_s\pi\eta a_b}{kT}\right)^{-0.67} \tag{7}$$

$$A_s = \frac{2(1-(1-p)^{1.67})}{2-3(1-p)+3(1-p)^{1.67}-2(1-p)^2} \tag{8}$$

with p being the porosity of column, a_b and a_s are the radii of bacteria and the glass beads, respectively, η and ρ_l are the absolute viscosity and density of PB buffer, μ is the approach velocity, g is the gravitational acceleration, k is the Boltzmann constant, T is the room temperature of 293 K. ρ_b is the density of the bacteria solution and A_{132} is the Hamaker constant as described by equation[65]:

$$A_{132} = (\sqrt{A_{11}} - \sqrt{A_{33}})(\sqrt{A_{22}} - \sqrt{A_{33}}) \tag{9}$$

A_{ii} denotes the individual Hamaker constant of bacteria (A_{11}), glass (A_{22}) and water (A_{33}), respectively. A_{33} was taken from literature[66] whereas A_{11} and A_{22} were obtained by equation[67]:

$$A_{ii} = 6\pi l_0^2\gamma_i^{LW} \tag{10}$$

According to Fowke[67], the value of $6\pi l_0^2$ equals 1.44×10^{-18} m^2, with l_0 being the minimum distance between the outermost cell surface and the glass bead (0.157 nm)[68].

The surface Gibbs free energies γ_i^{LW} were calculated according to Young's equation, based on measured contact angles (θ) of microbial lawns, and glass surfaces in three solvents (water, formamide and methylene iodide):

$$Cos(\theta) = -1 + 2\frac{\sqrt{\gamma_b^{LW}\gamma_l^{LW}}}{\gamma_l^{total}} + 2\frac{\sqrt{\gamma_b^+\gamma_l^-}}{\gamma_l^{total}} + 2\frac{\sqrt{\gamma_b^-\gamma_l^+}}{\gamma_l^{total}} \tag{11}$$

The total surface Gibbs free energies (γ^{total}) thereby were separated in a Lifshitz-van der Waals (γ^{LW}) and an acid-base component (γ^{AB}) with γ^+ and γ^- as the electron acceptor and the electron donor components of acid-base surface energy:

$$\gamma^{total} = \gamma^{AB} + \gamma^{LW} \tag{12}$$

$$\gamma_i^{AB} = 2\sqrt{\gamma_i^+\gamma_i^-} \tag{13}$$

Using literature data[38] of γ, γ^{LW}, γ^+, γ^- values for water, formamide, and methylene iodide, the parameters γ_b, γ_b^{LW}, γ_b^+, γ_b^- of bacteria were calculated as proposed by van Oss et al[69], and the data from literature were taken for assessing the free energy of the glass surface[70].

2.3.2 Electrokinetic Effects on Bacterial Deposition on Planar Surfaces

QCM-D is a high sensitivity method to measure the mass of bacterial attachment on a crystal sensor by observing the frequency decrease of crystal vibration (frequency shift) and the energy loss (dissipation energy shift) after cutting off energy supply[71,72]. The mass of attachment can be described by the Sauerbrey equation in case of rigid attachment[73]

$$\Delta f = \frac{-2f_0^2 \Delta m}{A\sqrt{\rho_q \mu_q}} = -C_f \Delta m \tag{14}$$

where f_0 denotes the fundamental resonance frequency, Δm is the mass of bacterial deposition, A is the electrode area, ρ_q is the density of quartz (2.648 g cm^{-3}) and μ_q is the shear modulus of quartz (2.957×1010 N/m^2). It should be noted that the mass of deposited bacteria can only be defined by this equation when rigid deposition (i.e., dissipation energy shift equals 0), otherwise the frequency shift needs to combine with microscope cell counting to quantify the mass of bacterial deposition. The dissipation energy shift versus the frequency shift (i.e., $\Delta f/\Delta D$) indicates energy dissipation per coupled unit mass, which gives the indications of the rigidity of bacterial adhesion[74–76]. In normal situations, bacterial adhesion leads to a negative frequency shift and positive dissipation energy shift. Thus less negative $\Delta f/\Delta D$ value indicates a dissipative soft and fluid film on the QCM-D sensor, and a more negative $\Delta f/\Delta D$ value stands for a more rigid layer[74]. These signals may thus help elucidate the rigidity and attachment strength of bacterial deposition.

Various electrolyte concentrations of 10, 50, and 100 mmol L^{-1} resulted in three different electrical double layer thicknesses, thus different electrokinetic effects, all of these factors were taken account for the net force quantification, to further investigate the electrokinetic-effect principles.

2.3.3 The Driving Force Dominating Bacterial Deposition and Transport

In both porous media and QCM-D planar systems, the same types of forces are acting on a bacterium in an external DC field: bacterial-solid surface interaction force (i.e., F_{DLVO}), hydraulic shear force (F_{HF}), electroosmotic shear force (F_{EOF}), and electrophoretic force (F_{EP}). The net force dominating bacterium deposition, therefore, can be expressed by the equation:

$$F_{net} = F_{DLVO} + F_{HF} + F_{EOF} + F_{EP} \tag{15}$$

The DLVO theory is a widely used approach that describes the interaction energy between bacteria and the solid surface combining electrostatic energy and Lifshitz-van der Waals interaction energy[69,77,78]

$$G_{\text{DLVO}} = G_{\text{EDL}} + G_{\text{LW}} \tag{16}$$

Previous research found that the strongest attractive energy which dominates the reversible bacterial deposition locates at the secondary minimum distance, bacterial cells deposit on the collector surface when their kinetic energy is lower than the attractive energy of the collector surface[79]. The DLVO attractive force dominating reversible bacterial deposition efficiency, therefore, can be calculated by the equation:

$$F_{\text{DLVO}} = \frac{G_{\text{DLVO}}}{h_s} \tag{17}$$

The shear forces F_{HF} and F_{EOF}, acting on a bacterium located at h_s depend on the velocities of the hydraulic (V_{HF}) and the electroosmotic (V_{EOF}) water flow and can be calculated by equations[80]

$$F_{\text{HF}} = F_d^* \times 6\pi\eta a V_{\text{HF}} \tag{18}$$

$$F_{\text{EOF}} = F_d^* \times 6\pi\eta a V_{\text{EOF}} \tag{19}$$

where η is the viscosity of the liquid, F_d^* is a function of the radius a of a sphere (for simplicity we presume bacterial cells to be spheres) and the distance of the center of the sphere to the collector surface. Following previous work we presume F_d^* to be 1.7[80].

The drag force F_{EP} acting on a bacterium is calculated from the electrophoretic velocity V_{EP} according to equation[81,82]

$$F_{\text{EP}} = 6\pi\eta a V_{\text{EP}} \tag{20}$$

As described previously, the electroosmosis and electrophoresis counteract each other to dominate the overall electrokinetic effects, the relative strength hence becomes the indicator to identify the dominating factor of the two forces. After eliminating the identical components electric field strength E, the permittivity of free space ε_0, dielectric constant of water ε_r, and reciprocal of viscosity $1/\eta$. The relative strength of $|F_{\text{EOF}}|$ and $|F_{\text{EP}}|$ can be simplified to the expression:

$$\frac{|F_{\text{EOF}}|}{|F_{\text{EP}}|} = \frac{F_d^* \zeta_C}{\frac{2}{3}\zeta_{\text{bac}} f(\kappa a)}[1 - \frac{2I_1(\kappa h_s)}{\kappa a I_0(\kappa h_s)}] \tag{21}$$

From the expression we understand that the force strength ratio is dominated by the electrical double layer thickness κ^{-1}, particle size a, and zeta potentials of matrix surface (ζ_C) and bacteria (ζ_{bac}). In particular, in 100 mM phosphate buffer $f(\kappa a)$ levels to 1.5, the function $[1-2I_1(\kappa h_s)/\kappa a I_0(\kappa h_s)]$ levels to 1, therefore the force ratio can be approximated to 1.29 $\zeta_C/\zeta_{\text{bac}}$. In this case, the dominating factor of the interactions between electroosmosis and electrophoresis can be simplified to the zeta potential ratio.

2.4 Electrokinetic Effects on Contaminant Sorption and Desorption

The advancing of the industry has significantly raised the overall life quality for human beings in the past two centuries, but at the same time, it has been inevitably accompanied by numerous side effects on the ecosystem. Polycyclic aromatic hydrocarbons (PAHs) are a group of typical contaminations widely distributed all over the world due to coal mining, oil exploitation, transport, leakage, and burning of oil, etc.[83,84]. They pose threat to human health due to persistent toxic and bioaccumulation proterties[85]. Massive PAH polluted soil areas have been found all over the world, especially the coal gasification and oil exploitation sites. Previous research has shown that PAH-degrading bacterial strains[86,87] are effective in decomposing contaminants[9]. However, the consensus in previous research is that only the contaminants dissolved in water are available to microbes[88]. The release of PAH contaminant to interact with live PAH-degraders in the ecosystem therefore plays an important role in determining PAH bioavailability.

Therefore, interactions with solid geo-sorbents are key drivers of the persistence of PAHs and, hence, control their fate as well as the exposure to environmental and human receptors. Various studies have shown that the sequestration of hydrophobic chemicals in the solid phase significantly reduce PAH bioavailability and biodegradation. Three potential rate-limiting steps may influence the sorption of contaminant to and its release from geo-matrices, respectively: (i) diffusion of the contaminant within the molecular nano-porous network, (ii) pore or surface diffusion in aggregated geo-matrices, and (iii) diffusion of the sorbate across an aqueous boundary layer surrounding sorbent particles. As a consequence of progressive binding, the residual hydrophobic contaminant may become less leachable and thus less efficiently available for microbial degradation[7]. The sorption and release of PAH in various sorbents can be investigated by kinetic[89–91] and thermodynamic (e.g., Gibbs free energy of sorption, ΔG^o) approaches[92,93].

2.4.1 Electrokinetic Effects on PHE Sorption/Desorption Kinetics

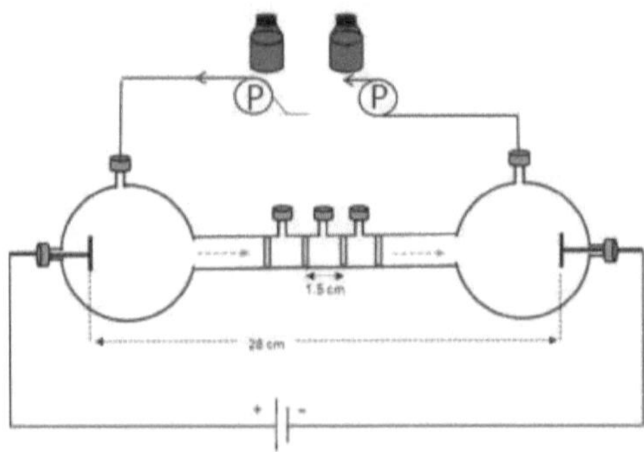

Figure 7. Schematic of the kinetic sorption/desorption set-up (adapted from Shan et al.[94])

Electrokinetic effects on PHE (as a typical PAH contaminant) sorption and desorption on mineral and carbonaceous matrices were investigated in a set-up as shown in Fig. 7. Breakthrough curves in adsorption are the course of the effluent adsorptive concentration at the outlet of fixed bed geo-matrices, it is an important approach for understanding the adsorption and releasing processes and the characterization of porous materials. In order to quantitatively describe the electrokinetic effects, normalized time-dependent fractions of PHE in the sorbent in desorption and sorption experiments were defined as $\Gamma_{des,t}$ (%) and $\Gamma_{sor,t}$ (%), respectively. They were calculated from PHE inflow (C_i) and outflow concentrations (C_e) of the reactor chamber, the electrolyte volume flushed (V; L), and the initial PHE load (M_0; mg) in the sorbent and the maximum amount of PHE that can be loaded on clean sorbent in the column (M_s; mg), respectively.

$$\Gamma_{des,t} = \frac{M_0 - \int_0^t C_e\, dV}{M_0} \tag{22}$$

$$\Gamma_{sor,t} = \frac{\int_0^t C_i\, dV - \int_0^t C_e\, dV}{M_s} \tag{23}$$

The relative influence of DC electric fields on PHE desorption ($\Delta\Gamma_{des,t}$) and sorption ($\Delta\Gamma_{sor,t}$) at a given time can be calculated by the following equations where subscripts denote the absence and presence of the electric field.

$$\Delta\Gamma_{des,t} = \Gamma_{des,noDC,t} - \Gamma_{des,DC,t} \tag{24}$$

$$\Delta\Gamma_{sor,t} = \Gamma_{sor,noDC,t} - \Gamma_{sor,DC,t} \tag{25}$$

2.4.2 The Relative Strength of Sorption and EOF Dominating Electrokinetic Effects

If geo-sorbents and contaminant solutions are allowed to interact long enough, equilibrium will be established between the amount of adsorbate adsorbed in solid and the amount of adsorbate in solution. Equilibrium is the most important piece of information for the understanding of how much adsorbate can be accommodated by a solid sorbent[95]. The equilibrium can be described by adsorption isotherms. The Freundlich equation is one of the earliest empirical equations used to describe equilibrium data and sorption characteristics for a heterogeneous surface[95,96]

$$\log q_e = \log K_F + n \log C_e \tag{26}$$

where q_e is the equilibrium concentration of PHE adsorbed to sorbents, C_e is the dissolved PHE equilibrium concentration, n is the Freundlich exponent (a measure of sorption linearity) and K_F is the Freundlich isotherm constant ($\mu g\ kg^{-1}$) ($L\ \mu g^{-1}$)^n. The distribution coefficient K_d at equilibrium was determined by $K_d = q_e/C_e$ ($L\ g^{-1}$). The specific surface-normalized distribution coefficient K_d^* can be further calculated by dividing K_d by the specific surface area ($m^2\ g^{-1}$) of the sorbents. Investigating the sorption equilibrium and isotherms at several different temperatures, thermodynamic parameters sorption Gibbs free energy, sorption enthalpy, and sorption entropy can be calculated. The Gibbs free energy of sorption ($\Delta G°$) relates to sorption enthalpy ($\Delta H°$) and sorption entropy changes ($\Delta S°$) by the equation:

$$\Delta G° = \Delta H° - T\Delta S° \tag{27}$$

$\Delta G°$ can be estimated according to the following equation:

$$\Delta G° = -RT \ln K_c \tag{28}$$

K_c is the equilibrium constant. It is dimensionless and based on the Freundlich isotherm K_F and the water density (ϱ) 1000 g L^{-1}, and can be calculated using equation[97–99]:

$$K_c = \frac{K_F \rho}{1000}\left(\frac{10^6}{\rho}\right)^{(1-n)} \tag{29}$$

$\Delta H°$ can be estimated using the Van't Hoff equation by substituting eq. 31 to eq. 30[97,99]:

$$lnK_c = \frac{-\Delta H°}{R} \times \frac{1}{T} + \frac{\Delta S°}{R} \tag{30}$$

The $\Delta H°$ (kJ mol^{-1}) is a measure of the enthalpy change (isosteric heat) involved in the transfer of solute from the reference state to the sorbed state at a given solid-phase concentration. R is the universal gas constant (8.314 × 10^{-3} kJ mol^{-1} K^{-1}) and T is the temperature in Kelvin. The values of $\Delta H°$ can be estimated by the slope and intercept of a plot of ln K_c versus $1/T$, and $\Delta S°$ can be calculated by $\Delta H°$ and $\Delta G°$ according to eq.27.

The Gibbs free energy change $\Delta G°$ indicates the degree of the spontaneity of PHE sorption on sorbents. Therefore, we challenged to apply $\Delta G°$ and the velocity of EOF to quantitatively correlate the relative strength of sorption and desorption factors, to find out the principles of electrokinetic effects.

2.5 Aims of this Study

Electrokinetics has been discussed as a promising technique to interfere with interactions of bacteria and contaminants with surfaces, and, hence, to improve contaminant bioavailability. However, the drivers of electrokinetic effects on bioavailability are not yet clear. This book hence strived (i) to quantify electrokinetic effects, (ii) to interlink observed effects to calculated electrokinetic forces, and (iii) to develop approaches to predict the electrokinetic effects on bacterial deposition and contaminant matrix interactions. In detail, it aimed

i)   to examine electrokinetic effects on bacterial transport in percolated laboratory columns in the presence and absence of electric fields and to develop an approach to predict electrokinetic effects based on bacterial and collector surface properties.

ii)   to investigate electrokinetic effects on bacterial deposition on flat surfaces by high accuracy QCM-D monitoring and microscopy approaches in the presence and absence of electric fields and to develop a predictive model for electrokinetic effects on bacterial deposition at varying electrolyte strengths.

iii)   to quantify electrokinetic effects on phenanthrene interactions (sorption/release) with mineral and carbonaceous sorbents in the presence and absence of electric fields and to develop an approach to predict electrokinetic effects on the sorption and desorption of contaminants based on the Gibbs free energy change of sorption (ΔG^{o}) and the EOF velocity.

3. Electrokinetic Effects on Bacterial Transport in Porous Media

3.1 Electric Field Effects on Bacterial Deposition and Transport in Porous Media

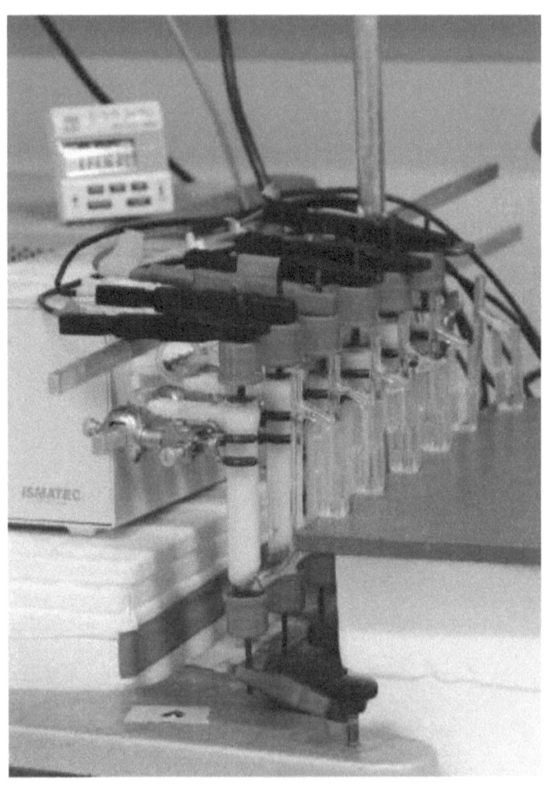

Cite This: *Environ. Sci. Technol.* 2018, 52, 14294−14301

pubs.acs.org/est

Electric Field Effects on Bacterial Deposition and Transport in Porous Media

Yongping Shan, Hauke Harms, and Lukas Y. Wick[*]

UFZ - Helmholtz Centre for Environmental Research, Department of Environmental Microbiology, Permoserstrasse 15, 04318 Leipzig, Germany

Ⓢ *Supporting Information*

ABSTRACT: Bacterial deposition and transport are key to microbial ecology and biotechnological applications. We therefore tested whether electrokinetic forces (electroosmotic shear force (F_{EOF}), electrophoretic drag force (F_{EP})) acting on bacteria may be used to control bacterial deposition during transport in laboratory percolation columns exposed to external direct current (DC) electric fields. For different bacteria, yet similar experimental conditions we observed that DC fields either enhanced or reduced bacterial deposition efficiencies (α) relative to DC-free controls. By calculating the DLVO force of colloidal interactions, F_{EOF}, F_{EP}, and the hydraulic shear forces acting on single cells at a collector surface we found that DC-induced changes of α correlated to $|F_{EOF}|$ to $|F_{EP}|$ ratios: If $|F_{EOF}| > |F_{EP}|$, α was clearly increased and if $|F_{EOF}| < |F_{EP}|$ α was clearly decreased. Our findings allow for better prediction of the forces acting on a bacterium at collector surface and, hence, the electrokinetic control of microbial deposition in natural and manmade ecosystems.

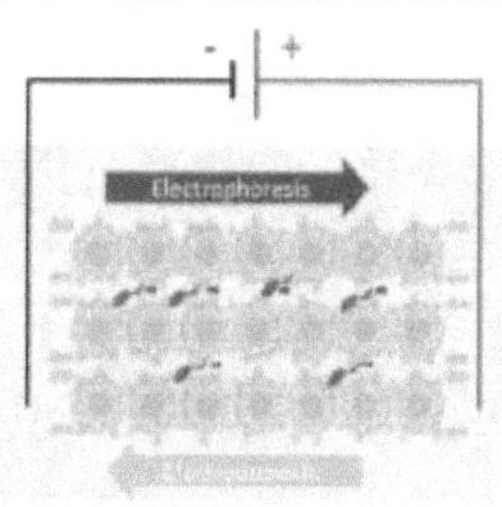

■ INTRODUCTION

Transport and deposition of bacteria are fundamental processes in microbial ecology and biotechnology.[1] They enable microbial functions in disturbed systems[2] or promote the formation of biofilms as a major life form of bacteria. While the catabolic activity of biofilms provide essential ecosystem services in natural and manmade systems (e.g., for the degradation of anthropogenic chemicals or in wastewater treatment), biofouling[3] by contrast may give rise to unwanted corrosion of metals,[4] clogging of filters/membranes or may even threaten human health by infecting medical devices[1] or technical systems for the provision of drinking water. There is, hence, strong interest in measures to control microbial deposition to surfaces as the first step in the formation of biofilms. Bacterial deposition is influenced by physicochemical properties of the microbe, the collector surface, and the aqueous medium.[5] Deposition to collector surfaces during transport in porous systems can be suitably approximated by the collision efficiency α, and clean-bed filtration theory,[6,7] while the distance-dependent energy between a bacterium and a collector surface (G_{DLVO}) can be quantified by the Derjaguin, Landau, Verwey, and Overbeek (DLVO) theory.[8] As the deposition of a bacterium requires that its kinetic energy is lower than its interaction energy with a collector surface, α, normally is positively correlated to G_{DLVO} at the distance of reversible attachment (i.e., at the secondary minimum of G_{DLVO}).[9] Although the DLVO theory refers to an ideal system (i.e., does not encompass heterogeneities in surface charge,[10,11] surface roughness,[12] hydration effects, or hydrophobic interaction[6,13]), it has been found to be a good predictor of bacterial deposition in solutions of high ionic strength (I = 0.1−0.3 M)[9,10,14] and/or to highly uniform surfaces of low surface roughness.

The electric field-induced phenomena electroosmosis and electrophoresis have been found to be powerful tools in controlling the movement of bacteria and (bio)colloidal particles.[15−18] When a DC electric field is applied to an ionic solution in a solid matrix, it invokes various electrokinetic transport processes: electromigration and electrophoresis denote the transport of charged molecules and particles, to the electrode of opposite charge, whereas electroosmotic flow (EOF) refers to the surface charge-induced movement of pore fluids usually from the anode to the cathode.[19] Due to its plug shape flow profile, EOF has been found to be efficient at a distance of a few nanometers above the solid surface, i.e., where bacteria-surface interactions affect the deposition efficiency of bacteria.[20] Both phenomena are directly correlated to the electric field strength applied; they allow for the movement of bacteria and colloidal particles[15−18] in porous media even in the absence of a pressure-driven hydraulic flow[21−23] or for the separation of monoclonal bacteria differing in the zeta potentials.[24]

Inspired by such observations, recent work analyzed the DLVO forces (F_{DLVO}), electroosmotic shear force (F_{EOF}), and hydraulic shear force (F_{HF}) acting on a bacterium at the secondary minimum distance and described F_{EOF} as a relevant

Received: July 3, 2018
Revised: November 2, 2018
Accepted: November 12, 2018
Published: November 12, 2018

DOI: 10.1021/acs.est.8b03648
Environ. Sci. Technol. 2018, 52, 14294−14301

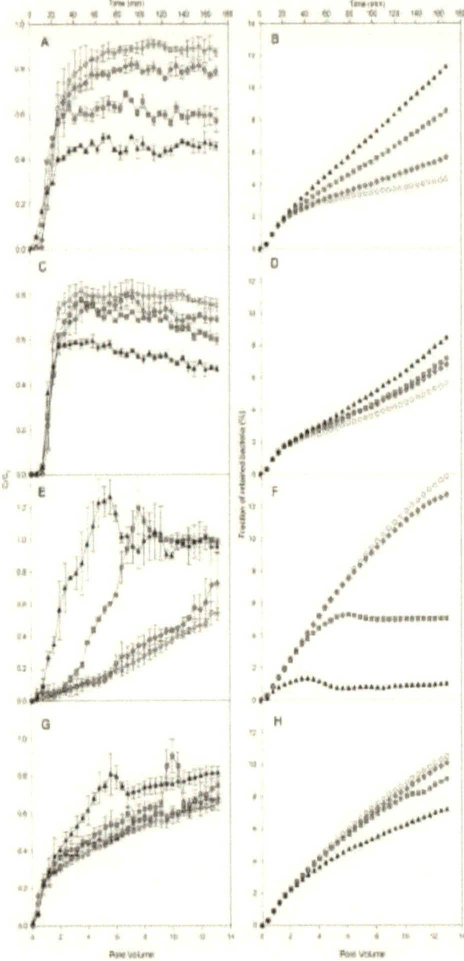

Figure 1. Breakthrough curves (left, data represent averages and standard deviations of triplicate experiments) and calculated fractions (right) of four bacteria transported through percolation columns packed with glass beads in the absence (open circle) and presence (filled symbols) of DC electric fields of $E = 1.0$ V cm^{-1} (rhomboids), $E = 2.0$ V cm^{-1} (squares) and $E = 3.0$ V cm^{-1} (triangles): *P. putida* KT2440 (A and B), *R. opacus* X9 (C and D), *P. fluorescens* LP6a (E and F), and *Sphingomonas* sp. S3 (G and H).

driver for the reduction of the initial adhesion of *Pseudomonas fluorescens* LP6a.[30] This approach, however neglected the electrophoretic drag force (F_{EP}) acting on bacteria and hence was unable to predict the interplay of F_{EOF}, F_{EP}, and F_{DLVO}. Here we experimentally quantify the effect of DC fields on the transport and deposition of four bacteria differing in their surface charge (zeta potential) and hydrophobicity in percolation columns. The observed DC field effects on bacterial deposition

efficiencies (α_t) relative to DC-free controls are reflected by calculations of the net force (F_{net}) acting on a bacterium at secondary minimum distance by F_{EOF}, F_{EP}, F_{HF}, and F_{DLVO}.

■ MATERIALS AND METHODS

Cultivation of Bacteria and Preparation of Inocula. *Pseudomonas putida* KT2440 (GenBank Accession No.

DOI: 10.1021/acs.est.8b03648
Environ. Sci. Technol. 2018, 52, 14294–14301

AE015451),[25] *Rhodococcus opacus* X9 (GenBank Accession No. AF095715),[26] *Pseudomonas fluorescens* LP6a (GenBank Accession No. AF525494)[27] and *Sphingomonas* species S3 (GenBank Accession No. MH048882) were cultivated in 500 mL Erlenmeyer flasks using 200 mL of minimal medium containing 1.0 g L^{-1} glucose (25 °C, rotary shaker at 150 rpm). The cultures were harvested in the early stationary phase (i.e., after 14 h for strain *P. putida* KT2440, 15 h for strain *R. opacus* X9, 12 h for strain *P. fluorescens* LP6a, and 7 d for strain *Sphingomonas* sp. S3, centrifuged at 3000g and resuspended in 100 mM potassium phosphate buffer (PB, pH 7, prepared by adding 0.061 mol K$_2$HPO$_4$ and 0.039 mol KH$_2$PO$_4$ in 1 L deionized water) with a Vortex mixer (Vortes-Genie 2, Scientific Industries, USA) to obtain an optical density of OD$_{600\ nm}$ = 0.30 using an UV/vis Spectrophotometer (Evolution 160, Thermo Fisher Scientific, Carlsbad, CA).

Characterization of Physico-Chemical Surface Properties of Bacteria and Glass Beads. The zeta potential (ζ) of bacteria and smashed glass beads were measured by Doppler electrophoretic light scattering analysis (Zetamaster, Malvern Instruments, Malvern, UK, with a Dip Cell Kit or a Folded Capillary Cell) in 100 mM PB (pH 7). In deviation from an earlier described procedure[20] analyses were performed at 60 V in order to obtain narrow and symmetrical signal peaks. To approximate the effect of bacterial deposition on the zeta potential of glass beads (0.1−0.25 mm diameter, Retsch, Germany), clean polished glass beads were smashed with a mortar and a pestle to a size of <100 μm, then heated at 200 °C in muffle furnace for 2 h, allowed to cool down to room temperature (25 °C) under sterile conditions and then immersed during 2 h to the bacterial suspensions (OD$_{600\ nm}$ = 0.30). The beads then were separated by sieving, rinsed cautiously with 100 mM PB, resuspended in 100 mM PB and analyzed as described above. Glass beads that were treated identically yet not exposed to bacterial cells were measured to obtain the ζ of the clean bed (i.e., collector) surfaces. The contact angles (Θ) of the bacteria were measured using a DSA 100 drop-shape analysis system (Krüss GmbH, Hamburg, Germany) using water (Θ_w), formamide (Θ_f), and methylene iodide (Θ_m) as described earlier.[20] Bacterial lawns were prepared by depositing bacteria from inoculated suspensions on cellulose acetate membrane filters (Millipore, 0.45 μm). Nine bacterial lawns were prepared for each bacterial cultivation to perform triplicate experiments for each solvent. Four droplets were applied on each bacterial lawn (i.e., the contact angle in each solvent is an average of 12 droplets). Glass bead lawns were prepared by fixing (either clean or bacteria-covered) glass beads with double-sided tape to glass slides by gentle pressing as described by Achtenhagen et al.[28] Glass beads of similar bacterial coverage as calculated for conditions of late stage breakthrough curves (cf. Tables 2 and S6 and Figures 1 and S2) were prepared as described in the SI. The contact angles of the glass beads are averages of 12 droplets).

Column Deposition Experiments. The breakthrough curves of the different strains were quantified in vertical percolation columns as described by Qin et al.[30] Shortly, the columns were sterilized and packed with clean, heat-sterilized (200 °C, 2 h) polished glass beads, the porosity and pore volume (PV) were estimated to be 0.42 and 3.97 mL, respectively. Two disk-shaped Ti/Ir electrodes (De Nora Deutschland GmbH, Germany) at the top (cathode) and bottom (anode) of the column were connected to a power pack (P333, Szczecin, Poland) that allowed to apply constant DC electric field at E = 0

(control), 0.5, 1.0, 1.5, 2.0, 2.5, or 3.0 V cm^{-1}. The columns were allowed to equilibrate by circulating clean buffer (100 mM PB, I = 0.22 M) for 30 min. Well stirred bacterial suspensions were allowed to percolate through the columns at a hydraulic flow rate of 19.3 mL h^{-1} (flow velocity: 2.4 × 10^{-7} m s^{-1}) from the top to the bottom using a peristaltic pump. By placing the anode at the outflow of the column potential impacts of anodic reactive oxygen species on bacterial deposition were avoided. The lid at the top of the column permitted the release of electrolytically formed gas bubbles and, hence, avoided the passage of gas bubbles through the packed bed. For some strains (*P. putida* KT2440, *P. fluorescens* LP6a) additional experiments with reversed electrode polarity (top: anode; bottom: cathode) at E = 2 V cm^{-1} were performed. The deposition of cells was determined by comparing the OD$_{600\ nm}$ of the influent (C_0) and effluent (C).

THEORY

Calculation of Collision Efficiency. Clean-bed filtration theory was used to quantify the bacterial deposition in the glass beads packed columns in the presence and absence of electric fields. The collision efficiency α_t is described by[6,7]

$$\alpha_t = \frac{\eta_t}{\eta_{train}} \tag{1}$$

with η_t being the rate of attachment as calculated from bacterial breakthrough data and η_{train} the rate of bacteria transport to the collector surfaces, the calculation method has been described in detail by Qin et al.[30] and in the Supporting Information (SI).

Prediction of Forces Acting on a Cell at the Secondary Minimum above a Collector Surface. According to the DLVO theory, the DLVO energy distribution (G_{DLVO}, eq 2)[8] is composed of the electrostatic (G_{EDL}) and Lifshitz−van der Waals (G_{LW}) energies (for detailed description please refer to the SI). The zeta potentials and contact angles of bacterial and collector surfaces, respectively, were used to approximate the overall DLVO interaction energies; Calculations of G_{DLVO} thereby considered changes of the zeta potential and the contact angles of the collector surface in response to increasing bacterial deposition (for detailed description of the effects of bacterial coverage on the zeta potential and contact angles of the collectors refer to the SI; eqs S9−S15, Table S6, and Figure S5).

$$G_{DLVO} = G_{EDL} + G_{LW} \tag{2}$$

At the secondary minimum distance (h_s) to a collector surface, F_{DLVO} (eq 3) can be calculated by the DLVO energy distribution (G_{DLVO}):

$$F_{DLVO} = \frac{G_{DLVO}}{h_s} \tag{3}$$

The resulting net force (F_{net}) acting on a bacterium located at the distance of the secondary minimum above a collector surface submersed in an ionic solution in the presence of an external DC electric field can be approximated by combination of shear forces induced by the hydraulic (F_{HF}), and the electroosmotic (F_{EOF}) water flow and the electrophoretic drag force (F_{EP}) in eq 4:

$$F_{net} = F_{EOF} + F_{EP} + F_{HF} + F_{DLVO} \tag{4}$$

The shear forces F_{HF} and F_{EOF}, acting on a bacterium located at h_s depend on the velocity of the hydraulic (V_{HF}) and the

DOI: 10.1021/acs.est.8b03546
Environ. Sci. Technol. 2018, 52, 14294−14301

Table 1. Overview of the Bacterial Zeta Potential and the Water Contact Angles, The Zeta Potential of the Collector Surface (Glass Beads) After Bacterial Deposition and the Calculated Clean Bed Deposition Efficiency (α_0; 0–2 PV) and the Deposition Efficiency (α_t) at Quasi Steady State of the Breakthrough Curves in Absence (no DC) and Presence of DC Electric Fields of Varying Field Strength ($E = 0$–3 V cm^{-1})[a]

bacteria name	zeta potential of bacteria ζ_{bac} (mV)	zeta potential of collector surface with bacteria[c] ζ_{C_b} (mV)	water contact angle Θ_w (degree)	water contact angle of collector with bacteria[c] $\Theta_{w,b}$ (degree)	collision efficiency (no DC) $\alpha_{no\,DC}$ / $\alpha_{no\,DC}$[d] (×10^{-2})	collision efficiency (1 V cm^{-1}) $\alpha_{0,1V/cm}$ / $\alpha_{0,1V/cm}$[d] (×10^{-2})	collision efficiency (2 V cm^{-1}) $\alpha_{0,2V/cm}$ / $\alpha_{0,2V/cm}$[d] (×10^{-2})	collision efficiency (3 V cm^{-1}) $\alpha_{0,3V/cm}$ / $\alpha_{0,3V/cm}$[d] (×10^{-2})
P. putida KT2440	−11 ± 1	−11 ± 3	70 ± 3	25 ± 3	28 (0.95) 0.44 ± 0.04[e]	25 (0.87) 0.88 ± 0.07[e]	19 (0.78) 1.89 ± 0.17[e]	19 (0.84) 3.01 ± 0.13[e]
R. opacus X9	−18 ± 3	−15 ± 2	62 ± 3	30 ± 2	43 (0.94) 1.01 ± 0.12[f]	34 (0.90) 1.34 ± 0.26[f]	24 (0.88) 1.55 ± 0.20[f]	15 (0.91) 1.68 ± 0.21[f]
P. fluorescens LP6a[b]	−35 ± 3[c]	−16 ± 3	46 ± 3	34 ± 5	26 (0.98) 1.7 ± 0.16[f,g]	19 (0.83) 0.94 ± 0.19[f,g]	19 (0.98) 0.28 ± 0.03[f,g]	4 (0.95) 0 ± 0[f,g]
Sphingomonas sp. S3	−23 ± 2	−15 ± 4	53 ± 5	39 ± 4	38 (0.67) 1.65 ± 0.34[h]	17 (0.63) 1.48 ± 0.36[h]	19 (0.81) 1.21 ± 0.29[h]	12 (0.89) 0.98 ± 0.21[h]

[a]Please note that quasi steady state of the breakthrough curves is reached at different times for the bacteria analyzed as specified in the footnote to this table. [b]data taken from ref 20. [c]The ζ_C and Θ_w of clean glass bead collectors were 8 ± 1 mV and 21 ± 2°, respectively (cf. SI Table S6). [d]The values in brackets refer to the coefficient of determination r^2. [e]Calculated as average from 5–13 PV. [f]Calculated as average from 5–13 PV. [g]Calculated as average from 20–25 PV (cf. SI Figure S2). [h]Calculated as average from 8–13 PV.

electroosmotic (V_{EOF}) water flow and can be calculated by eqs 5 and 6:[29]

$$F_{HF} = F_d^* \times 6\pi\eta a V_{HF} \tag{5}$$

$$F_{EOF} = F_d^* \times 6\pi\eta a V_{EOF} \tag{6}$$

Where η is the viscosity of the liquid ($\eta = 3.19$ kg m^{-1} h^{-1}), F_d^* is a function of the radius a of a sphere (for simplicity we presume bacterial cells to be spheres) and the distance of the center of the sphere to the collector surface. Following previous work we presume F_d^* to be 1.7.[29] The velocity of hydraulic flow can be described by the Hagen–Poiseuille approach.[30] The EOF velocity (V_{EOF}) at distance h_s from the collector surface is calculated by eq 7, which is the combination of a simplified EOF expression of the Navier–Stokes equation with the potential distribution described by the Gouy–Chapman model, and the characteristics of porous media were taken into account.[30,31,32]

$$V_{EOF} = -\frac{\varepsilon_0 \varepsilon_r \zeta_C n\tau \times E}{\eta}\left(1 - \frac{2I_1(\kappa h_s)}{\kappa a I_0(\kappa h_s)}\right) \tag{7}$$

In eq 7 ε_r is the dielectric constant of water (78.5), ε_0 (8.85 × 10^{-12} F m^{-1}) is the vacuum permittivity, ζ_C is the zeta potential of the collector surface at the experimental conditions, n and τ refer to the porosity (0.42) and tortuosity (1.8) of the glass bead bed,[32] and E is the electric field strength applied, I_0 and I_1 are the zero- and first-order modified Bessel functions, and κ^{-1} is the thickness of the electric double layer. The drag force F_{EP} acting on a bacterium is calculated from the electrophoretic mobility V_{EP} according to Solomentsev et al.[30,33]

$$F_{EP} = 6\pi\eta a V_{EP} \tag{8}$$

$$V_{EP} = \frac{2\varepsilon_0 \varepsilon_r \zeta_{bac} E}{3\eta} f_{(\kappa a)} \tag{9}$$

The electrophoretic velocity (V_{EP}) is calculated by the Smoluchowski equation.[34] $f(\kappa a)$ approaches 1 for small κa, and 1.5 for large κa, here $f(\kappa a)$ value level off to 1.5.[33] The ratio of F_{EOF} and F_{EP} is given by eq 10

$$\frac{F_{EOF}}{F_{EP}} = \frac{F_d \times \zeta_C n\tau}{\frac{2}{3} \times \zeta_{bac} f(\kappa a)}\left(1 - \frac{2I_1(\kappa h_s)}{\kappa a I_0(\kappa h_s)}\right) = 1.29 \times \frac{\zeta_C}{\zeta_{bac}} \tag{10}$$

RESULTS

Quantification of Cell Deposition in Percolation Columns. The effects of DC electric fields on bacterial deposition and transport of P. putida KT2440, R. opacus X9, P. fluorescens LP6a, and Sphingomonas sp. S3 were quantified in percolation columns filled with polished glass beads at various electric field strengths ($E = 0$–3.0 V cm^{-1}). By quantifying relative effluent cell densities, the breakthrough curves of DC and DC-free columns were compared and clean bed theory was adopted to describe the bacterial deposition. The collision efficiency of the clean bed (i.e., at the initial stage of bacterial breakthrough; α_0) was evaluated from data of 0–2 PV of the breakthrough curves (Table 1, Figure 1 and SI Figures S1 and S3), while later stage collision efficiencies (α_t) were obtained from the breakthrough curves at quasi steady state (Table 1, Figures 1 and S2). All four bacterial strains differed in their physicochemical surface properties. In the percolation buffer they exhibited zeta potentials (ζ_{bac}) of −11 to −35 mV (Table 1). All strains were moderately hydrophobic with water contact angles varying between 46° and 70°. Such differences were also reflected by distinct breakthrough curves in DC-free controls: strains KT2440 and X9 were less retained than strains LP6a and S3 (Table 1); this is reflected by smaller collision efficiencies ($\alpha_t \approx 0.004$–0.01 vs 0.02; Table 1) and lower fractions of retained bacteria after 14 PV ($\approx 4\%$ vs 10–14%, Figure 1). No significant differences of the clean bed collision efficiencies ($\alpha_0 \approx 0.3$–0.4), however, were calculated. The zeta potential of the glass beads

DOI: 10.1021/acs.est.8b03440
 Environ. Sci. Technol. 2018, 52, 14294–14301

Environmental Science & Technology

Article

Table 2. Overview of Forces Acting on a Bacterium at the Distance of the Secondary Minimum for Deposition to a Clean Bed (0–2 PV; Denominated by the Subscript "0") and at Quasi Steady State of the Breakthrough Curves (Denominated by the Subscript "t") in Presence and Absence of DC Electric Fields of Varying Field Strength ($E = 0–3$ V cm^{-1}): DLVO Interaction Force (F_{DLVO}), Electroosmotic Shear Force (F_{EOF}), Electrophoretic Drag Force (F_{EP}), the Hydraulic Shear Force (F_{HF}) and the Net Force (F_{net}) According to eq 4

bacteria name	DLVO force at distance of 2nd minimum F_{DLVO_0} F_{DLVO_t} (pN)	electroosmotic shear force (per V cm^{-1} electric field strength) F_{EOF_0} F_{EOF_t} (pN)	electrophoretic drag force (per V cm^{-1} electric field strength) F_{EP} (pN)	hydraulic flow shear force F_{HF} (pN)	net force at distance of 2nd minimum (no DC) $F_{net\_0, no\ DC}$ $F_{net\_t, no\ DC}$ (pN)	net force at distance of 2nd minimum (1 V cm^{-1}) $F_{net\_0, 1\ V/cm}$ $F_{net\_t, 1\ V/cm}$ (pN)	net force at distance of 2nd minimum (2 V cm^{-1}) $F_{net\_0, 2\ V/cm}$ $F_{net\_t, 2\ V/cm}$ (pN)	net force at distance of 2nd minimum (3 V cm^{-1}) $F_{net\_0, 3\ V/cm}$ $F_{net\_t, 3\ V/cm}$ (pN)
P. putida KT2440	3.26 3.69[a]	1.36 1.87[a,b]	−1.45	0.2	3.06 3.49[a,b]	2.97 3.91[a,c]	2.88 4.33[a,c]	2.79 4.75[a,b]
R. opacus X9	5.61 7.62[a]	1.36 2.55[a,c]	−2.37	0.2	5.41 7.42[a,c]	4.4 7.6[a,c]	3.39 7.78[a,c]	2.38 7.96[a,c]
P. fluorescens LP6a	2.31 1.83[a]	1.36 2.72[a,d]	−4.74	0.2	2.11 1.63[a,d]	−1.27 −0.39[a,d]	−4.65 −2.41[a,d]	−8.03 −4.43[a,d]
Sphingomonas sp. S3	8.19 9.82[a]	1.36 2.55[a,e]	−3.03	0.2	7.99 9.62[a,e]	6.32 9.14[a,e]	4.65 8.66[a,e]	2.98 8.18[a,e]

[a]Calculated using respective ζ_{C_t} and contact angles of bacteria and bacteria adhered glass beads (cf. SI Tables S6). [b]Calculated based on ζ_{C_t} as average from 5–13 PV. [c]Calculated based on ζ_{C_t} as average from 5–13 PV. [d]Calculated based on ζ_{C_t} as average from 20–25 PV (cf. SI Figure S2). [e]Calculated based on ζ_{C_t} as average from 8–13 PV.

(ζ_C) changed from −8 mV (clean bed) to ca. −11 to −16 mV (ζ_{C_t}, Table 1) in response to bacterial deposition (surface coverage of 4–14% (SI Table S6)). Bacterial deposition likewise changed the contact angle of clean glass beads ($\Theta_w = 21°$) to 25° (P. putida KT2440), 30° (R. opacus X9), 34° (P. fluorescens LP6a) and 39° (Sphingomonas sp. S3) (Tables 1 and S6).

Applying DC fields to the columns resulted in changed breakthrough of all four strains. Observed effects depended on the electric field strengths applied and the zeta potential of the bacterial (ζ_{bac}) and the glass bead surfaces (ζ_C and ζ_{C_t}): At $\zeta_{bac}/\zeta_C \gtrsim 1.29$ the DC fields led to a decreased bacterial deposition, while at $\zeta_{bac}/\zeta_C \lesssim 1.29$ DC fields promoted bacterial deposition to the glass collector surfaces. Both the positive and negative DC effects on bacterial deposition increased at augmenting field strengths (Table 1 and Figure 1 for $E = 0$, 1, 2, and 3 V cm^{-1}; SI Figure S1 and Table S2 for $E = 0.5$, 1.5, and 2.5 V cm^{-1}). Applying DC fields to the columns hence decreased bacterial initial deposition for all strains at <2 PV. This resulted, for instance, in decreases of α_0 at $E = 3$ V cm^{-1} by 85% (LP6a), 68% (S3), 65% (X9), and 32% (KT2440) relative to DC-free controls. At >2 PV bacterial breakthrough showed two distinct tendencies: the presence of DC clearly decreased deposition of strains S3 and LP6a, as exemplified by 40–100% reduced α_t at $E = 3$ V cm^{-1} (Tables 1 and S2, Figures 1 and S1). By contrast, up to 584% increased collision efficiencies α_t of strains KT2440 and X9 were observed at $E = 3$ V cm^{-1} (Table 1, Figure 1).

Net Forces Acting on a Cell Placed at the Distance of the Secondary Minimum. Tables 2 ($E = 0, 1, 2, 3$ V cm^{-1}) and S3 ($E = 0, 0.5, 1.5, 2.5$ V cm^{-1}) summarize the net forces (F_{net}, cf. eq 4) acting on bacteria placed at the distance of the secondary minimum above a collector surface in presence and absence of DC electric fields of varying field strengths; that is, the DLVO force (F_{DLVO}, cf. eq 3), the hydraulic (F_{HF}, cf. eq 5) and electroosmotic shear forces (F_{EOF}, cf. eq 6 and 7), and the electrophoretic drag force (F_{EP}, cf. eq 8 and 9). The F_{DLVO} ranged from 1.83 to 9.82 pN and, the distance of the secondary minimum (cf. SI Figure S5 and Table S6), were significantly higher than strain-independent $F_{HF} = 0.2$ pN. The DLVO approach was used as it has been found to be a powerful predictor of bacterial deposition to chemically uniform collector surfaces of poor surface roughness immersed in solutions of high ionic strength ($I = 0.1–0.3$ M), that is, conditions as given in our experiments.

As both the bacterial and the glass collector surfaces were measured to be negatively charged (Table 1), the electrophoretic drag forces F_{EP} and the electroosmotic shear force counteracted each other (as expressed by opposite signs of the forces) and depended on the electric field strength applied. For situations of clean bed surfaces ($\zeta_C = -8$ mV), $|F_{EOF}| < |F_{EP}|$ was calculated for all strains and all conditions tested (Table 2). In such situation and in the presence of DC F_{net} was consistently $< F_{DLVO}$ and allowed for less bacterial deposition (i.e., decreasing α_0 or better bacterial transport) than in DC-free controls. Due to the deposition-induced increase of the zeta potential (ζ_{C_t}) of the collector surfaces, the velocity of the EOF (V_{EOF}, eq 7) increased during percolation. This resulted in an increased F_{EOF} yet let F_{EP} unchanged. For strains LP6a and S3 $|F_{EOF}|$ remained $< |F_{EP}|$ while for strains KT2440 and X9 $|F_{EOF}|$ became $> |F_{EP}|$. As a consequence F_{net} remained $< F_{DLVO}$ for strains LP6a and S3 yet at $E = 3$ V cm^{-1} increased by 29% and 5% for strains KT2440 and X9 relative to F_{DLVO} (Tables 2 and S3).

■ DISCUSSION

Drivers of Deposition Efficiencies and Net Forces at the Secondary Minimum. Inspired by previous work[10] that interlinked reduced deposition efficiencies of P. fluorescens LP6a cells with electroosmotic shear forces (F_{EOF}) in electrokinetic percolation columns we here challenged the proposed F_{EOF}-

DOI: 10.1021/acs.est.8b03940
 Environ. Sci. Technol. 2018, 52, 14294−14301 effects by quantifying deposition and transport of four soil bacteria differing in their physicochemical cell surfaces and deposition properties. While our data confirm deposition-limiting F_{EOF}-effects for LP6a cells (i.e., that F_{EOF} are able to overcome the F_{DLVO}), we simultaneously found that DC electric fields promoted the deposition of *P. putida* KT2440 and *R. opacus* X9 up to 584% and 66% despite of $|F_{EOF}| \geq |F_{DLVO}|$ (Table 2). In order to explain such discrepancy we included the electrophoretic drag force, F_{EP} as additional driver of the F_{net} (eq 4) acting on a cell sitting at the distance of the secondary minimum (eq 4) above a glass bead collector surface. We found that the relative changes of DC-induced net forces (expressed by $(F_{net,DC} - F_{net,noDC})/F_{net,noDC}$) were highly correlated with the relative changes of the collision efficiency (expressed by $(a_{DC} - a_{noDC})/a_{noDC}$) (Figure 2). At conditions of $F_{net,DC} > F_{net,noDC}$

reduced deposition $(a_{DC} < a_{noDC})$ and at $1.29 \times \zeta_C/\zeta_{bac} > 1$ increased deposition $(a_{DC} > a_{noDC})$ is to be predicted (Figure 3).

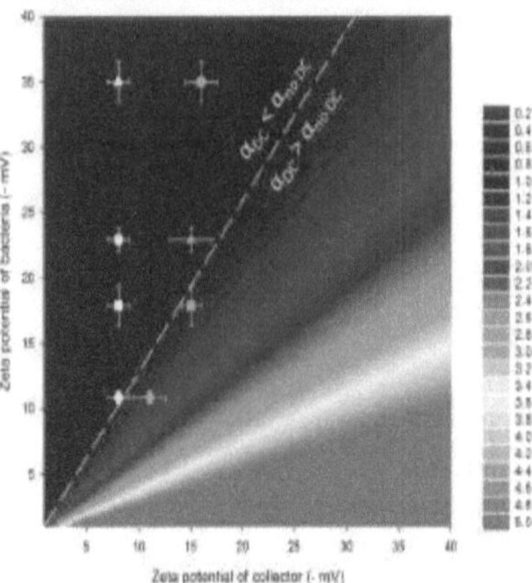

Figure 3. Calculated effects of the zeta potential of collector (ζ_C) and bacterial (ζ_{bac}) surfaces on $|F_{EOF}/F_{EP}|$ ratios (cf. eq 10). At $|F_{EOF}/F_{EP}| > 1$ increased and at $|F_{EOF}/F_{EP}| < 1$ decreased deposition of cells relative to DC-free controls, respectively, is expected. Open and gray filled symbols represent the averages and the standard error ($n = 3$) of the bacterial surfaces and the initial and late stage zeta potential of glass beads covered by *P. putida* KT2440 (diamonds), *R. opacus* X9 (squares), *P. fluorescens* LP6a (circles), and *Sphingomonas* sp. S3 (triangles). Differences of zeta potential of clean glass beads and glass beads covered with bacteria are statistically significant ($p < 0.05$) for all bacterial strains.

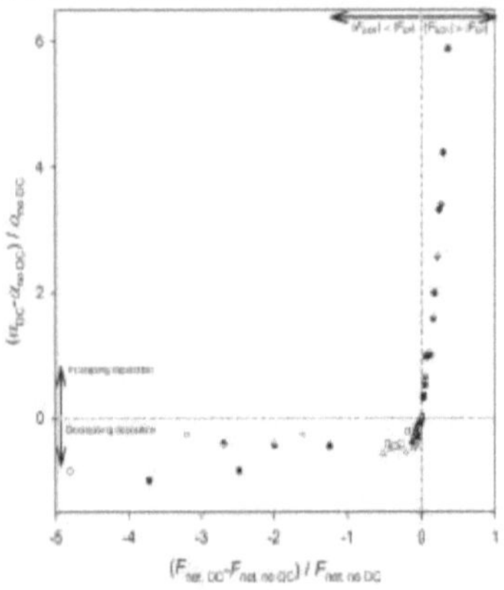

Figure 2. Relative changes of DC-induced net forces acting on a bacterium placed at the secondary and relative changes of the collision efficiency of *P. putida* KT2440 (diamonds), *R. opacus* X9 (squares), *P. fluorescens* LP6a (circles) and *Sphingomonas* sp. S3 (triangles) cells. Open and filled symbols represent relative changes for deposition to clean beds (0–2 PV) and at quasi steady state stages of the breakthrough curves (cf. Table 2). Semifilled symbols represent relative changes in the presence of DC fields with reversed polarity applied (i.e., allowing for EOF in direction of the hydraulic flow); top-filled and bottom-filled symbols refer to for deposition to clean beds (0–2 PV) and at quasi steady state stages of the breakthrough curves (cf. Table 2).

(i.e., $|F_{EOF}| > |F_{EP}|$) increased deposition, while for $F_{net,DC} < F_{net,noDC}$ (i.e., $|F_{EOF}| < |F_{EP}|$) decreased deposition was detected for all bacteria and all stages of the breakthrough curves. Comparison of the absolute values of F_{EOF} and F_{EP} (Table 2) reveals that at $|F_{EOF}| > |F_{EP}|$ improved and at $|F_{EOF}| < |F_{EP}|$ reduced bacterial deposition relative to DC-free controls was observed. Our data hence suggest that electrokinetic shear and drag forces are drivers of electrokinetic influences on bacterial deposition. As both forces are influenced by the same drivers (e.g., electric field strength and the thickness of the electric double layer), the $|F_{EOF}/F_{EP}|$ ratio can be expressed in a given medium in function of the zeta potentials of the bacteria and the collector surface by $1.29 \times \zeta_C/\zeta_{bac}$ (eq 10): at $1.29 \times \zeta_C/\zeta_{bac} < 1$

The zeta potential ratio between collector and bacteria hence seems to be key to electrokinetic control of bacterial deposition and transport. Although other descriptors exist, we used the zeta potentials that were derived from electrophoretic mobility measurements using the standard Smoluchowski theory. Such approach has been described to be adequate and to be a better predictor for bacterial deposition rates than outer surface potentials described by soft particle theory.[36]

We challenged our observations by reversing the direction of the electric field during deposition experiments of *P. putida* KT2440 and *P. fluorescens* LP6a (SI Figure S4, Tables S4 and S5). As expected, the resulting F_{net} differed due to changed relative direction of the EOF and the hydraulic flow in the columns. However, no significant changes in the overall deposition and transport during breakthrough of the two bacterial strains were observed (Figure S4) with data fitting well to Figure 2.

Relevance for Environmental and Biotechnological Application. Our results suggest that DC electric field effects may be used to control bacterial deposition and transport in immersed porous media given the properties and heterogeneities of the collector and bacterial surfaces. Such effects may help to increase the retention of unwanted bacteria in drinking water purification systems[37,38] or, vice versa, to reduce biofouling[39–41] and biocorrosion[42] in technical systems. Our data further show that both F_{EOF} and F_{EP} act on bacterial deposition at extents depending on the $|F_{EOF}/F_{EP}|$ and ζ_C/ζ_{bac}

Environmental Science & Technology

ratios, respectively (Figure 3). Increasing the surface charge of the collector supports the deposition of bacteria (or any other colloid) and may promote desired biofilm formation (e.g., in water cleanup systems), while reduction of ζ_C/ζ_{bac} reduces bacterial deposition and, hence, biofilm formation in technical systems where it is undesired (Figure 3). In our study shifts of ζ_C coincided with electrokinetically induced changes of the deposition efficiency due to priming of the collector surface to more negative zeta potentials (Table 1); such priming due to continuous deposition of bacteria during percolation has also been described earlier.[20] Tailor-made (i.e., dynamic and possibly reversible) changes of ζ_C hence may be applied in technical applications in order to find solutions for the wanted ζ_C/ζ_{bac} ratios and bacterial deposition, respectively. Drivers of zeta potential variation such as material properties, ionic strength, and pH then become available to steer electric field effects to the aimed direction. For instance, priming of the collector (e.g., with highly charged materials or solutes) will support the deposition of bacteria (or any other colloid) and promote wanted biofilm formation. In drinking water purification systems, with typical low ionic strength (<10 mM[45]) and neutral pH, the collector matrices (i.e., ion-exchange resins, activated carbon, etc.) typically are highly charged (ζ_C of ca. -50 mV), and high ζ_C/ζ_{bac} ratios are relatively easy to achieve. This may lead to increased bacterial deposition, increased removal of microbial pathogens and, hence, a promotion of drinking water safety. On the other hand, reduction of ζ_C/ζ_{bac} reduces bacterial deposition and hence, biofilm formation in technical systems (Figure 3). Several DC-based approaches have been proposed to influence bacteria-electrode surface interactions: some studies aimed at disrupting biofilm formation on electrodes by applying a biocidal current[20] while others used electrokinetic approaches for better application to biofilm.[36,44] Weak DC electric fields have not been found to negatively affect bacterial physiology and activity,[20,45] nor to change bacterial physicochemical surface properties relevant for adhesion and transport.[20] Applying DC fields also opens possibilities for enhanced bacterial transport in porous natural matrices. Investigations have found that in the natural soil system where typical zeta potential distribution ranges of bacteria (-5 to -48 mV[46,47]) and matrices (0 to -54 mV[48–50]) are relatively wide, the two different effects of electric fields exist at the same time regarding the ζ_C/ζ_{bac} distribution. For the situations $\zeta_C/\zeta_{bac} > 1.29$ (i.e., $|F_{EOF}| > |F_{EP}|$), DC fields enhance the deposition of bacterial in porous matrices, however, the strong F_{EOF} may enhance the desorption and migration of contaminants,[51] and thus may also bridge the physical distance between bacterium and contaminants to further enhance bioremediation. On the other hand, at $\zeta_C/\zeta_{bac} < 1.29$ (i.e., $|F_{EOF}| < |F_{EP}|$), DC fields may enhance the transport of bacteria through porous media to reach contaminants adsorbed on matrices, and enhance bioremediation. In electrokinetically managed natural and manmade ecosystems knowledge of the electroosmotic flow and electrophoresis hence allows for better control of microbial deposition transport in porous media.

■ ASSOCIATED CONTENT

⬡ Supporting Information

The Supporting Information is available free of charge on the ACS Publications website at DOI: 10.1021/acs.est.8b03648.

Text, six tables, and five figures (PDF)

■ AUTHOR INFORMATION

Corresponding Author

*Phone: +49 341 235 1316; fax: +49 341 235 45 1316; e-mail: lukas.wick@ufz.de.

ORCID

Lukas Y. Wick: 0000-0001-7296-865X

Notes

The authors declare no competing financial interest.

■ ACKNOWLEDGMENTS

This work has been supported by Helmholtz Centre for Environmental Research (UFZ) within the 'Chemicals in the Environment' research topic and the China Scholarship Council (CSC). The authors wish to thank Jana Reichenbach, Rita Remer, and Birgit Würz for wonderful and skilled technical support and Jinyi Qin (Chang'an University; China) on the advices on the experimental design. Provision of strain *Sphingomonas* sp. S3 by Nelson Khan (UFZ and University of Nairobi; Kenya) is greatly acknowledged.

■ REFERENCES

(1) Gavin, L.; Gillian, D. L. *Microbial Biofilms: Current Research and Applications*; Caister Academic: Norfolk, U.K., 2012.

(2) Worrch, A.; König, S.; Banitz, T.; Centler, F.; Frank, K.; Thullner, M.; Harms, H.; Miltner, A.; Wick, L. Y.; Kästner, M. Bacterial Dispersal Promotes Biodegradation in Heterogeneous Systems Exposed to Osmotic Stress. *Front. Microbiol.* **2016**, 7, No. 1214.

(3) Meng, F.; Zhang, S.; Oh, Y.; Zhou, Z.; Shin, H.-S.; Chae, S.-R. Fouling in Membrane Bioreactors: An Updated Review. *Water Res.* **2017**, *114*, 151–180.

(4) Parker, C. Species of Sulphur Bacteria Associated with the Corrosion of Concrete. *Nature* **1947**, *159* (4039), 439–440.

(5) Torkzaban, S.; Bradford, S. A.; Walker, S. L. Resolving the Coupled Effects of Hydrodynamics and DLVO Forces on Colloid Attachment in Porous Media. *Langmuir* **2007**, *23* (19), 9652–9660.

(6) Martin, R. E.; Bouwer, E. J.; Hanna, L. M. Application of Clean-Bed Filtration Theory to Bacterial Deposition in Porous Media. *Environ. Sci. Technol.* **1992**, *26* (5), 1053–1058.

(7) Velasco-Casal, P.; Wick, L. Y.; Ortega-Calvo, J.-J. Chemoeffectors Decrease the Deposition of Chemotactic Bacteria during Transport in Porous Media. *Environ. Sci. Technol.* **2008**, *42* (4), 1131–1137.

(8) *Particle Deposition and Aggregation: Measurement, Modelling and Simulation*; Elimelech, M., Ed.; Colloid and surface engineering series; Butterworth-Heinemann: Oxford, 1998.

(9) Redman, J. A.; Walker, S. L.; Elimelech, M. Bacterial Adhesion and Transport in Porous Media: Role of the Secondary Energy Minimum. *Environ. Sci. Technol.* **2004**, *38* (6), 1777–1785.

(10) Song, L.; Johnson, P. R.; Elimelech, M. Kinetics of Colloid Deposition onto Heterogeneously Charged Surfaces in Porous Media. *Environ. Sci. Technol.* **1994**, *28* (6), 1164–1171.

(11) Elimelech, M.; Chen, J. Y.; Kuznar, Z. A. Particle Deposition onto Solid Surfaces with Micropatterned Charge Heterogeneity: The "Hydrodynamic Bump" Effect. *Langmuir* **2003**, *19* (17), 6594–6597.

(12) Bhattacharjee, S.; Ko, C.-H.; Elimelech, M. DLVO Interaction between Rough Surfaces. *Langmuir* **1998**, *14* (12), 3365–3375.

(13) Hermansson, M. The DLVO Theory in Microbial Adhesion. *Colloids Surf., B* **1999**, *14* (1–4), 105–119.

(14) Simoni, S. F.; Bosma, T. N. P.; Harms, H.; Zehnder, A. J. B. Bivalent Cations Increase Both the Subpopulation of Adhering Bacteria and Their Adhesion Efficiency in Sand Columns. *Environ. Sci. Technol.* **2000**, *34* (6), 1011–1017.

(15) Hu, Y.; Werner, C.; Li, D. Electrokinetic Transport through Rough Microchannels. *Anal. Chem.* **2003**, *75* (21), 5747–5758.

(16) Masliyah, J. H.; Bhattacharjee, S. *Electrokinetic and Colloid Transport Phenomena*; John Wiley & Sons: NJ, 2006.

Environmental Science & Technology

Article

(17) Pennathur, S.; Santiago, J. G. Electrokinetic Transport in Nanochannels. 1. Theory. *Anal. Chem.* **2005**, *77* (21), 6772–6781.

(18) Kuo, C.-C.; Papadopoulos, K. D. Electrokinetic Movement of Settled Spherical Particles in Fine Capillaries. *Environ. Sci. Technol.* **1996**, *30* (4), 1176–1179.

(19) Kirby, B. J.; Hasselbrink, E. F. Zeta Potential of Microfluidic Substrates: 1. Theory, Experimental Techniques, and Effects on Separations. *Electrophoresis* **2004**, *25* (2), 187–202.

(20) Qin, J.; Sun, X.; Liu, Y.; Berthold, T.; Harms, H.; Wick, L. Y. Electrokinetic Control of Bacterial Deposition and Transport. *Environ. Sci. Technol.* **2015**, *49* (9), 5663–5671.

(21) DeFlaun, M. F.; Condee, C. W. Electrokinetic Transport of Bacteria. *J. Hazard. Mater.* **1997**, *55* (1–3), 263–277.

(22) Secord, E. L.; Kottara, A.; Van Cappellen, P.; Lima, A. T. Inoculating Bacteria into Polycyclic Aromatic Hydrocarbon-Contaminated Oil Sands Soil by Means of Electrokinetics. *Water, Air, Soil Pollut.* **2016**, *227* (8), 288.

(23) Haber, S. Deep Electrophoretic Penetration and Deposition of Ceramic Particles inside Impermeable Porous Substrates. *J. Colloid Interface Sci.* **1996**, *179* (2), 380–390.

(24) Wick, L. Y.; Mattle, P. A.; Wattiau, P.; Harms, H. Electrokinetic Transport of PAH-Degrading Bacteria in Model Aquifers and Soil. *Environ. Sci. Technol.* **2004**, *38* (17), 4596–4602.

(25) Nelson, K. E.; Weinel, C.; Paulsen, I. T.; Dodson, R. J.; Hilbert, H.; Martins dos Santos, V. A. P.; Fouts, D. E.; Gill, S. R.; Pop, M.; Holmes, M. Complete Genome Sequence and Comparative Analysis of the Metabolically Versatile *Pseudomonas putida* KT2440. *Environ. Microbiol.* **2002**, *4* (12), 799–808.

(26) Furuno, S.; Remer, R.; Chatzinotas, A.; Harms, H.; Wick, L. Y. Use of Mycelia as Paths for the Isolation of Contaminant-Degrading Bacteria from Soil: Use of Mycelia as Paths for the Isolation of Bacteria. *Microb. Biotechnol.* **2012**, *5* (1), 142–148.

(27) Foght, J. M.; Westlake, D. W. Transposon and Spontaneous Deletion Mutants of Plasmid-Borne Genes Encoding Polycyclic Aromatic Hydrocarbon Degradation by a Strain of *Pseudomonas fluorescens*. *Biodegradation* **1996**, *7* (4), 353–366.

(28) Achtenhagen, J.; Goebel, M.-O.; Miltner, A.; Woche, S. K.; Kästner, M. Bacterial Impact on the Wetting Properties of Soil Minerals. *Biogeochemistry* **2015**, *122* (2–3), 269–280.

(29) Goldman, A. J.; Cox, R. G.; Brenner, H. Slow Viscous Motion of a Sphere Parallel to a Plane Wall—II Couette Flow. *Chem. Eng. Sci.* **1967**, *22* (4), 653–660.

(30) Probstein, R. F. *Physicochemical Hydrodynamics: An Introduction*; John Wiley & Sons, 2005.

(31) Ghosal, S. Fluid Mechanics of Electroosmotic Flow and Its Effect on Band Broadening in Capillary Electrophoresis. *Electrophoresis* **2004**, *25* (2), 214–228.

(32) Shi, L.; Müller, S.; Harms, H.; Wick, L. Y. Factors Influencing the Electrokinetic Dispersion of PAH-Degrading Bacteria in a Laboratory Model Aquifer. *Appl. Microbiol. Biotechnol.* **2008**, *80* (3), 507–515.

(33) Solomentsev, Y.; Böhmer, M.; Anderson, J. L. Particle Clustering and Pattern Formation during Electrophoretic Deposition: A Hydrodynamic Model. *Langmuir* **1997**, *13* (23), 6058–6068.

(34) Locke, B. R. Electrophoretic Transport in Porous Media: A Volume-Averaging Approach. *Ind. Eng. Chem. Res.* **1998**, *37* (2), 615–625.

(35) Hunter, R. J. *Zeta Potential in Colloid Science: Principles and Applications*, 3rd printing; Colloid Science; Academic Press: London, 1988.

(36) de Kerchove, Alexis J.; Elimelech, Menachem Relevance of Electrokinetic Theory for "Soft" Particles to Bacterial Cells: Implications for Bacterial Adhesion. *Langmuir* **2005**, *21* (14), 6462–6472.

(37) Besra, L.; Liu, M. A Review on Fundamentals and Applications of Electrophoretic Deposition (EPD). *Prog. Mater. Sci.* **2007**, *52* (1), 1–61.

(38) T, P. A.; Rolf, B.; J, B. H. Controlled Electrophoretic Deposition of Bacteria to Surfaces for the Design of Biofilms. *Biotechnol. Bioeng.* **2000**, *67* (1), 117–120.

(39) Zumbusch, P. v; Kulcke, W.; Brunner, G. Use of Alternating Electrical Fields as Anti-Fouling Strategy in Ultrafiltration of Biological Suspensions – Introduction of a New Experimental Procedure for Crossflow Filtration. *J. Membr. Sci.* **1998**, *142* (1), 75–86.

(40) Jagannadh, S. N.; Muralidhara, H. S. Electrokinetics Methods To Control Membrane Fouling. *Ind. Eng. Chem. Res.* **1996**, *35* (4), 1133–1140.

(41) G, B.; E, O. Reduction of Membrane Fouling by Means of an Electric Field During Ultrafiltration of Protein Solutions. *Berichte Bunsenges. Phys. Chem.* **1989**, *93* (9), 1026–1032.

(42) Lin, J.; Ballim, R. Biocorrosion Control: Current Strategies and Promising Alternatives. *Afr. J. Biotechnol.* **2012**, *11* (91), 15736–15747.

(43) *Guidelines for Drinking-Water Quality*, 4th ed.; World Health Organization: Geneva, 2011.

(44) *Encyclopedia of Membranes*; Drioli, E., Giorno, L., Eds.; Springer Berlin Heidelberg: Berlin, Heidelberg, 2016.

(45) Shi, L.; Müller, S.; Loffhagen, N.; Harms, H.; Wick, L. Y. Activity and Viability of Polycyclic Aromatic Hydrocarbon-degrading Sphingomonas Sp. LB126 in a DC-electrical Field Typical for Electrobioremediation Measures. *Microb. Biotechnol.* **2008**, *1* (1), 53–61.

(46) Soni, K. A.; Balasubramanian, A. K.; Beskok, A.; Pillai, S. D. Zeta Potential of Selected Bacteria in Drinking Water When Dead, Starved, or Exposed to Minimal and Rich Culture Media. *Curr. Microbiol.* **2008**, *56* (1), 93–97.

(47) Van Loosdrecht, M. C.; Lyklema, J.; Norde, W.; Schraa, G.; Zehnder, A. J. Electrophoretic mobility and hydrophobicity as a measured to predict the initial steps of bacterial adhesion. *Appl. Environ. Microbiol.* **1987**, *53* (8), 1898–1901.

(48) Stenström, T. A. Bacterial hydrophobicity, an overall parameter for the measurement of adhesion potential to soil particles. *Appl. Environ. Microbiol.* **1989**, *55* (1), 142–147.

(49) Vane, L. M.; Zang, G. M. Effect of Aqueous Phase Properties on Clay Particle Zeta Potential and Electro-Osmotic Permeability: Implications for Electro-Kinetic Soil Remediation Processes. *J. Hazard. Mater.* **1997**, *55* (1–3), 1–22.

(50) Yukselen, Y.; Kaya, A. Zeta Potential of Kaolinite in the Presence of Alkali, Alkaline Earth and Hydrolyzable Metal Ions. *Water, Air, Soil Pollut.* **2003**, *145* (1–4), 155–168.

(51) Wick, L. Y.; Shi, L.; Harms, H. Electro-Bioremediation of Hydrophobic Organic Soil-Contaminants: A Review of Fundamental Interactions. *Electrochim. Acta* **2007**, *52* (10), 3441–3448.

DOI: 10.1021/acs.est.8b03648
Environ. Sci. Technol. 2018, 52, 14294–14301

3.2 Supporting Information

Electric Field Effects on Bacterial Deposition and Transport in Porous Media

Yongping Shan, Hauke Harms, and Lukas Y. Wick*

UFZ - Helmholtz Centre for Environmental Research, Department of Environmental Microbiology, 04318 Leipzig, Germany

Number of pages: 15

Number of figures: 5

Number of tables: 6

*Corresponding author: Mailing address: Helmholtz Centre for Environmental Research - UFZ. Department of Environmental Microbiology; Permoserstrasse 15; 04318 Leipzig, Germany. phone: +49 341 235 1316, fax: +49 341 235 45 1316, e-mail: lukas.wick@ufz.de.

CALCULATIONS

Please note that the calculation of the collision efficiency, the surface coverage, the fraction of bacteria retained, and the DLVO energy profile have been published in the Supporting information of a previous work in our group (DOI: 10.1021/es506245y):

Collision efficiency (α_t)

The collision efficiency (α_t) of bacteria was calculated applying the clean-bed filtration theory [1] (eq. S2). The α_t of bacteria is defined as the ratio of the rate of attachment (η_t) to the rate of bacterial transport to the surfaces (η_{trans})

$$\alpha_t = \frac{\eta_t}{\eta_{trans}} \tag{S1}$$

Values of η_{trans} were calculated taking into account the contributions of convection, diffusion, van der Waals attraction, and sedimentation [2]. For the calculations, we assumed spheres of identical size glass beads (diameter: 0.1 mm) in their closest packing, and identical effective bacterial radius (1 μm) of the bacteria. Values of η_t were calculated from C/C_0 values obtained in column experiments.

$$C = C_0 \exp\left(-\frac{3(1-\varepsilon)}{4a_s}\eta_{trans}\alpha_t L\right) \tag{S2}$$

where C is the effluent cell concentration, C_0 the influent cell concentration, ε the porosity of the packed bed, a_s the radius of the glass beads, L the length of the column, and η_t the transport of bacteria from the solution to the glass surface in the whole experimental time. η_{trans} was approximated by applying the solution to convection-diffusion equation (eq. S3):

$$\eta_{trans} = A_s \frac{A_{132}}{9\pi\eta a_b^2 \mu}^{0.125} \frac{a_b}{a_s}^{1.875} + 0.00338 A_s \frac{2a_b^2(\rho_b - \rho_l)g}{9\eta\mu}^{1.2} \frac{a_b}{a_s}^{-0.4} + 4A_s^{0.33}\frac{12\mu a_s \pi\eta a_b}{kt}^{-0.67} \tag{S3}$$

$$A_s = \frac{2(1-(1-\varepsilon)^{1.67})}{2-3(1-\varepsilon)+3(1-\varepsilon)^{1.67}-2(1-\varepsilon)^2} \tag{S4}$$

with ε being the porosity of column (0.41), a_b and a_s are the radii of bacteria ($a_b = 10^{-6}$ m) and the glass beads ($a_s = 5 \times 10^{-5}$ m), respectively, η and ρ_l are the absolute viscosity ($\eta = 3.19$ kg m^{-1} h^{-1}) and density of PB buffer ($\rho = 1018$ kg m^{-3}), μ is the approach velocity (1.1×10^{-4} m s^{-1}), g is the gravitational acceleration (9.81 m s^{-2}), k is the Boltzmann constant (1.38×10^{-23} J

K^{-1}), t is the room temperature of 293 K. ρ_b is the density of the bacteria solution . ($\rho_b = 1090$ $kg\ m^{-3}$) and A_{132} is the Hamaker constant [2] as described by eq. S5 [3]

$$A_{132} = (\sqrt{A_{11}} - \sqrt{A_{33}})(\sqrt{A_{22}} - \sqrt{A_{33}}) \tag{S5}$$

Here, A_{ii} denotes the individual Hamaker constant of bacteria (A_{11}), glass (A_{22},) and water (A_{33}), respectively. A_{33} was taken from literature [4] whereas A_{11} and A_{22} were obtained by eq. S6 [5].

$$A_i = 6\pi l_0^2 \gamma_i^{LW} \tag{S6}$$

According to Fowkes [5], the value of $6\pi l_0^2$ equals 1.44×10^{-18} m^2, with l_0 being the minimum distance between the outermost cell surface and the glass bead (0.157 nm) [6].

Surface coverage and the fraction of bacteria retained

Assuming that the entire glass surface allows for irreversible adhesion, the fraction of bacterial coverage on the surface can be described by eq. S7:

$$\theta = \frac{N_b \pi a_b^2}{N_s 4\pi a_s^2} \tag{S7}$$

where N_s and N_b are the numbers of collectors and bacteria in the column and a_b and a_s the radii of bacteria and the collectors, respectively [8].

The fraction of bacteria retained in the column, R is calculated by eq. S8:

$$R = \int (1 - \frac{C}{C_0}) \cdot dV \tag{S8}$$

where C and C_0 are the effluent and influent cell concentrations of the column, and V the flow of the cell suspension ($V = 19$ ml h^{-1}) through the column [9, 10].

DLVO interaction energy of bacterial adhesion

According to the DLVO theory [11], the DLVO interaction energy of bacterial adhesion (G_{DLVO}) the electrostatic repulsion (G_{EDL}), and the Lifshitz-van der Waals (G_{LW}) energy (eq. S9) [11]:

$$G_{DLVO} = G_{EDL} + G_{LW} \tag{S9}$$

The surface Gibbs free energies of bacteria γ_b and the glass surface γ_s (mJ m^{-2}) were calculated based on measured contact angles (θ) of microbial lawns, membrane filters and glass surfaces using water, formamide and methylene iodide as liquids using the Young equation according to eq. S10:

$$Cos(\theta) = -1 + 2\frac{\sqrt{\gamma_b^{LW}\gamma_l^{LW}}}{\gamma_l^{total}} + 2\frac{\sqrt{\gamma_b^{+}\gamma_l^{-}}}{\gamma_l^{total}} + 2\frac{\sqrt{\gamma_b^{-}\gamma_l^{+}}}{\gamma_l^{total}} \tag{S10}$$

The total surface Gibbs free energies (γ^{total}) thereby were separated in a Lifshitz-van der Waals (γ^{LW}) and an acid-base component (γ^{AB}) (eq. S11) with γ^+ and γ^- as the electron acceptor and the electron donor components of acid-base surface energy (eqs. S11 & 12).

$$\gamma^{total} = \gamma^{AB} + \gamma^{LW} \tag{S11}$$

$$\gamma_l^{AB} = 2\sqrt{\gamma_l^{+}\gamma_l^{-}} \tag{S12}$$

Using literature data [14] of γ, γ^{LW}, γ^+, γ^- values for water, formamide and methyleneiodide, the parameters γ_b, γ_b^{LW}, γ_b^{+}, γ_b^{-} of bacteria were calculated as proposed by van Oss et al[15], and the data from literature taken for assessing the free energy of the glass surface [16].

Electrostatic repulsion energy (G_{EDL})

The electrostatic repulsion energy between bacteria and the glass surface was calculated by eq. S13 [11,12]:

$$G_{EDL} = \pi\varepsilon_0\varepsilon_r a_b \left\{ 2\xi_b\xi_s \ln\left[\frac{1+\exp(-\kappa h)}{1-\exp(-\kappa h)}\right] + (\xi_b^{2} + \xi_s^{2})\ln\left[1-\exp(-2\kappa h)\right] \right\} \tag{S13}$$

where κ^{-1} is the thickness of electrical double layer (EDL, nm) as calculated by the Guoy-Chapman theory with C and z being the molar bulk concentration and the charge number of the electrolytes[14] (eq. S14).

$$\kappa^{-1} = \left[3.29 z C^{1/2}\right]^{-1} \tag{S14}$$

For a 10 mM and a 100 mM PB solution, a κ^{-1} of 2.15 nm (10 mM PB) and κ^{-1} of 0.65 nm (100 mM PB) were calculated.

Lifshitz-van der Waals interaction energy (G_{LW})

With given values of the effective Hamaker constant, the Lifshitz-van der Waals interaction energy can be calculated by eq. S15 [13-15]

$$G_{LW} = -\frac{A_{132}}{6}\left[\frac{2a_b(h+a_b)}{h(h+2a_b)} - \ln(\frac{h+2a_b}{h})\right] \tag{S15}$$

Estimation of the effect of bacterial adhesion on changes of the physico-chemical surface properties (contact angle and zeta potential) of glass bead collectors

The coverage of bacteria on glass surface was estimated using eq. S16,

$$Coverage = \frac{\pi r_{bac}^2 \times N_{Cell}}{4\pi r_{glass}^2} \tag{S16}$$

where N_{cell} is the cell number deposited on glass surface, r_{bac} the radius of bacteria, r_{glass} the radius of crushed glass beads (0.01 mm). N_{cell} was calculated from changes of the OD at 600 nm (OD_{600nm}) of bacterial cell suspensions before after 2 h immersion of crushed glass beads (diameter of ca. 0.01 mm) in bacterial cell suspensions (100 mM phosphate buffer; initial OD_{600nm} = 0.30). In order to obtain similar bacterial coverage of the glass bead surface as calculated for late-stage of the breakthrough curves (cf. Tables 2 & S6) the following liquid to solid ratios (mL g^{-1}) were used: 3.2 mL g^{-1} for *P. putida* KT2440, 5.5 mL g^{-1} for *R. opacus* X9, 11 mL g^{-1} for *P. putida* LP6a and 7.6 mL g^{-1} for *Sphingomonas* sp. S3 respectively. The contact angles and zeta potentials of the glass beads were measured as described in the main text.

The effect of changing liquid to solid ratios (mL g^{-1}) on surface coverage (obtained by variations of the OD_{600nm} of the cell suspensions) and resulting contact angles and zeta potentials of the glass beads have been further studied and is exemplified in Fig. S5 for *P. putida* KT2440 and *P. fluorescens* LP6a, respectively. The data of the zeta potential changes of *P. putida* LP6a are in good agreement with previous findings published by Qin et al[17].

Table S1. Glossary of parameters used in calculations

Parameter	Description	Value taken from
C	the effluent cell concentration	own measurements
C_0	the influent cell concentration	own measurements
a_t	the radius of the glass beads	manufacturer's information
ε	the porosity of the packed bed	own measurements
L	the length of the column	own measurements
a_b	the radii of bacteria	own measurements
η	the absolute viscosity ($\eta = 3.19$ kg m^{-1} h^{-1})	literature cited in the text
ρ_l	density of PB buffer ($\rho = 1018$ kg m^{-3})	calculation
μ	the approach velocity (1.1×10^{-4} m s^{-1})	own measurements
g	the gravitational acceleration (9.81 m s^{-2})	constant
k	the Bolzmann constant (1.38×10^{-23} J K^{-1}),	constant
t	the room temperature of 293 K	constant
ρ_b	the density of the bacteria solution	literature cited in the text
A_{132}	the Hamaker constant	constant
A_{11}	individual Hamaker constant of bacteria	calculations with contact angles
A_{22}	individual Hamaker constant of glass	calculations with contact angles
A_{33}	individual Hamaker constant of water	calculations with contact angles
l_0	the minimum distance between the outermost cell surface and the glass bead (0.157 nm)	literature cited in the text
γ	surface free energy	calculations with contact angles
Θ_w	contact angles of water	own measurements
Θ_m	contact angles of methylene iodide	own measurements
Θ_f	contact angles of formamide	own measurements

Table S2. Overview of calculated clean bed deposition efficiency (α_0; 0 - 2 PV) and the deposition efficiency (α_t) at quasi steady state of the breakthrough curves in absence (no DC) and presence of DC electric fields of varying field strengths (E = 0.5, 1.5 & 2.5 V cm^{-1}).

Bacteria	collision efficiency (no DC)	collision efficiency (0.5 V cm^{-1})	collision efficiency (1.5 V cm^{-1})	collision efficiency (2.5 V cm^{-1})
	$\alpha_{0,no\,DC}$ [a] $\alpha_{t,no\,DC}$	$\alpha_{0,\,0.5\,V/cm}$ [a] $\alpha_{t,\,0.5\,V/cm}$	$\alpha_{0,\,1.5\,V/cm}$ [a] $\alpha_{t,\,1.5\,V/cm}$	$\alpha_{0,\,2.5\,V/cm}$ [a] $\alpha_{t,\,2.5\,V/cm}$
	($\times10^{-2}$)	($\times10^{-2}$)	($\times10^{-2}$)	($\times10^{-2}$)
P. putida KT2440	28 (0.95) 0.44 ± 0.04[b]	22 (0.88) 0.64 ± 0.04[b]	21 (0.91) 1.3 ± 0.15[b]	16 (0.87) 2.3 ± 0.07[b]
R. opacus X9	43 (0.94) 1.01 ± 0.12[c]	34 (0.90) 1.34 ± 0.26[c]	24 (0.88) 1.55 ± 0.20[c]	15 (0.91) 1.68 ± 0.21[c]
P. fluorescens LP6a [a]	26 (0.98) 1.7 ± 0.16[d]	n.a. [e]	n.a. [e]	n.a.[e]
Sphingomonas sp. S3	38 (0.67) 1.65 ± 0.34[f]	20 (0.68) 1.8 ± 0.12[f]	21 (0.71) 1.5 ± 0.19[f]	16 (0.66) 0.93 ± 0.09[f]

[a] the values in brackets refer to the coefficient of determination r^2; [b] calculated as average from 5-13 PV; [c] calculated average from 5 - 13 PV; [d] calculated average from 20 - 25 PV (cf. Fig S2); [e] not analysed; [f] calculated average from 8 - 13 PV.

Table S3. Overview of forces acting on a bacterium at the distance of the secondary minimum for deposition to a clean bed (0 - 2 PV; denominated by the subscript '0') and at quasi steady state of the breakthrough curves (denominated by the subscript 't') in presence and absence of DC electric fields of varying field strength ($E = 0.5$, 1.5 & 2.5 V cm^{-1}): DLVO interaction force (F_{DLVO}), electroosmotic shear force (F_{EOF}), electrophoretic drag force (F_{EP}), the hydraulic shear force (F_{HF}) and the net force (F_{net}) according to eq. 4 in main text.

Bacteria	DLVO force at distance of 2nd minimum	electroosmotic shear force (per V cm^{-1} electric field strength)	electrophoretic drag force (per V cm^{-1} electric field strength)	hydraulic flow shear force	net force at distance of 2nd minimum (no DC)	net force at distance of 2nd minimum (0.5 V cm^{-1})	net force at distance of 2nd minimum (1.5 V cm^{-1})	net force at distance of 2nd minimum (2.5 V cm^{-1})
	F_{DLVO_0} (F_{DLVO_t})	F_{EOF_0} F_{EOF_t}	F_{EP}	F_{HF}	$F_{net\_0, no DC}$ $F_{net\_t, no DC}$	$F_{net\_0, 0.5 V/cm}$ $F_{net\_t, 0.5 V/cm}$	$F_{net\_0, 1.5 V/cm}$ $F_{net\_t, 1.5 V/cm}$	$F_{net\_0, 2.5 V/cm}$ $F_{net\_t, 2.5 V/cm}$
	(pN)	(pN)	(pN)	(pN)	(pN)	(pN)	(pN)	(pN)
P. putida KT2440	3.26 3.69[a]	1.36 1.87[a,b]	-1.45	0.2	3.06 3.49[a,b]	3.02 3.70[a,b]	2.93 4.12[a,b]	2.84 4.54[a,b]
R. opacus X9	5.61 7.62[a]	1.36 2.55[a,c]	-2.37	0.2	5.41 7.42[a,c]	4.91 7.51[a,c]	3.90 7.69[a,c]	2.89 7.87[a,c]
P. fluorescens LP6a	2.31 1.83[d]	1.36 2.72[a,d]	-4.74	0.2	2.11 1.63[a,d]	0.42 0.62[a,d]	-2.96 -1.40[a,d]	-6.34 -3.42[a,d]
Sphingomonas sp. S3	8.19 9.82[d]	1.36 2.55[a,e]	-3.03	0.2	7.99 9.62[a,e]	7.16 9.38[a,e]	5.49 8.90[a,e]	3.82 8.42[a,e]

[a] calculated using respective $\zeta_{C,t}$; [b] calculated based on $\zeta_{C,t}$ as average from 5 - 13 PV; [c] calculated based on $\zeta_{C,t}$ as average from 5 - 13 PV; [d] calculated based on $\zeta_{C,t}$ as calculated based on $\zeta_{C,t}$ as average from 20 - 25 PV (cf. Fig S2); [e] calculated based on $\zeta_{C,t}$ as average from 8 - 13 PV.

Table S4. Overview of the calculated clean bed deposition efficiency (α_0; 0 - 2 PV) in absence (no DC) and presence of DC electric fields of varying field strength ($E = 0 - 3$ V cm^{-1}) of both electric field directions.

Bacteria	collision efficiency (no DC) $\alpha_{0,no\,DC}$[b] $\alpha_{1,no\,DC}$[c] ($\times10^{-2}$)	collision efficiency (1 V cm^{-1}) $\alpha_{0,1.0\,V/cm}$[b] $\alpha_{1,1.0\,V/cm}$ ($\times10^{-2}$)	collision efficiency (2 V cm^{-1}) $\alpha_{0,2.0\,V/cm}$[b] $\alpha_{1,2.0\,V/cm}$ ($\times10^{-2}$)	collision efficiency (3 V cm^{-1}) $\alpha_{0,3.0\,V/cm}$[b] $\alpha_{1,3.0\,V/cm}$ ($\times10^{-2}$)
P. putida KT2440	28 (0.95) 0.44 ± 0.04 [c]	25 (0.87) 0.88 ± 0.07 [c]	19 (0.78) 1.89 ± 0.17 [c]	19 (0.84) 3.01 ± 0.13 [c]
Reversed polarity [a] *P. putida* KT2440	28 (0.95) 0.44 ± 0.04 [c]	25 (0.95) 0.92 ± 0.03 [c]	20 (0.90) 1.53 ± 0.06 [c]	18 (0.72) 1.89 ± 0.09 [c]
P. fluorescens LP6a	26 (0.98) 1.7 ± 0.16 [d]	19 (0.83) 0.94 ± 0.19 [d]	19 (0.98) 0.28 ± 0.03 [d]	4 (0.95) 0 ± 0 [d]
Reversed polarity [a] *P. fluorescens* LP6a	26 (0.98) 1.7 ± 0.16 [d]	N.A. [e] N.A. [e]	15 (0.88) [d] 0.25 ± 0.06 [d]	N.A. [e] 0 ± 0 [d]

[a] reversed polarity refers to the electric field direction: top (anode) and bottom (cathode), while the other experiments were performed in electric field direction: top (cathode) and bottom (anode);

[b] the values in brackets refer to the coefficient of determination r^2; [c] calculated as average from 5 - 13 PV; [d] calculated as average from 20 - 25 PV; [e] data not available.

Table S5. Overview of forces acting on a bacterium at the distance of the secondary minimum for deposition to a clean bed (0 - 2 PV: denominated by the subscript '0') and at quasi-steady state of the breakthrough curves (denominated by the subscript 't') in presence and absence of DC electric fields of varying field strength ($E = 0 - 3$ V cm^{-1}): DLVO interaction force (F_{DLVO}), electroosmotic shear force (F_{EOF}), electrophoretic drag force (F_{EP}), the hydraulic shear force (F_{HF}) and the net force (F_{net}) according to eq. 4 in the main text.

Bacteria name	DLVO force at distance of 2nd minimum	electroosmotic shear force (per V cm^{-1} electric field strength)	electrophoretic drag force (per V cm^{-1} electric field strength)	hydraulic flow shear force	net force at distance of 2nd minimum (no DC)	net force at distance of 2nd minimum (1 V cm^{-1})	net force at distance of 2nd minimum (2 V cm^{-1})	net force at distance of 2nd minimum (3 V cm^{-1})
	F_{DLVO_0} (F_{DLVO_t})	F_{EOF_0} (F_{EOF_t})	F_{EP}	F_{HF}	$F_{net\_0, no DC}$ $F_{net\_t, no DC}$	$F_{net\_0, 1 V/cm}$ $F_{net\_t, 1 V/cm}$	$F_{net\_0, 2 V/cm}$ $F_{net\_t, 2 V/cm}$	$F_{net\_0, 3 V/cm}$ $F_{net\_t, 3 V/cm}$
	(pN)	(pN)	(pN)	(pN)	(pN)	(pN)	(pN)	(pN)
P. putida KT2440	3.26 3.69 [a]	1.36 1.87 [a,b]	-1.45	-0.2	3.06 3.49 [a,b]	2.97 3.91 [a,b]	2.88 4.33 [a,b]	2.79 4.75 [a,b]
Reversed polarity P. putida KT2440	3.26 3.69 [a]	1.36 1.87 [a,b]	-1.45	0.2	3.46 3.89 [a,b]	3.37 4.31 [a,b]	3.28 4.73 [a,b]	3.19 5.15 [a,b]
P. fluorescens LP6a	2.31 1.83 [a]	1.36 2.72 [a,c]	-4.74	-0.2	2.11 1.63 [a,c]	-1.27 -0.39 [a,c]	-4.65 -2.41 [a,c]	-8.03 -4.43 [a,c]
Reversed polarity P. fluorescens LP6a	2.31 1.83 [a]	1.36 2.72 [a,c]	-4.74	0.2	2.51 2.03 [a,c]	-0.87 0.01 [a,c]	-4.25 -2.01 [a,c]	-7.63 -4.03 [a,c]

[a] calculated using respective ζ_{C_t}; [b] calculated based on ζ_{C_t} as average from 5 - 13 PV; [c] calculated based on ζ_{C_t} as average from 20 - 25 PV.

Table S6. Contact angles of bacteria, clean glass beads and bacterial adhered glass beads, and the Hamaker constants (Aii) of bacteria (A_{11}), glass (A_{22}), and effective Hamaker constant (A_{132}) calculated based on the measured contact angles.

Bacteria	Coverage		Contact angle (Θ)			Hamaker constant				Secondary minimum distance	
	bacteria glass in batch	bacteria glass in column	Θ_w	Θ_f	Θ_m	A_{11}	A_{22}	A_{132} clean collector surface	A_{132} collector surface w/ bacteria [a]	clean collector surface	collector surface w/ bacteria
	(%)	(%)	(degree)					($*10^{-21}$ J)		(nm)	(nm)
P. putida KT2440			70 ± 3	64 ± 7	57± 2	44.1		0.31	0.37	3.2	
R. opacus X9			62 ± 3	45± 5	51 ± 4	48.2		0.52	0.78	3.2	
P. fluorescens LP6a			46 ± 3	55 ± 4	56 ± 2	45.6		0.34	0.34	4.1	
Sphingomonas sp. S3			53 ± 11	67 ± 2	45 ± 2	53.5		0.72	1.01	3.1	
Clean glass beads			21 ± 2	40 ± 5	56 ± 4	N.A.	44.5	N.A.	N.A.		
P. putida KT2440 covered glass beads (late stage)	5.3	2.9 - 4.4	25 ± 3	40 ± 1	54 ± 2	N.A.	46.1	N.A.	N.A.		3.3
R. opacus X9 covered glass beads (late stage)	7.3	2.5 - 5.8	30 ± 2	46 ± 3	51 ± 2	N.A.	48.6	N.A.	N.A.		3.5
P. fluorescens LP6a covered glass beads (late stage)	11.7	5.3 -14	34 ± 5	51 ± 6	56 ± 2	N.A.	44.5	N.A.	N.A.		4.7
Sphingomonas sp. S3 covered glass beads (late stage)	9.6	7.8 -10.7	39 ± 4	54 ± 3	55 ± 4	N.A.	42.8	N.A.	N.A.		3.4

a) values of the Hamaker constant were calculated using the physico-chemical surface properties of measured contact angles of bacterial covered glass beads in the table.

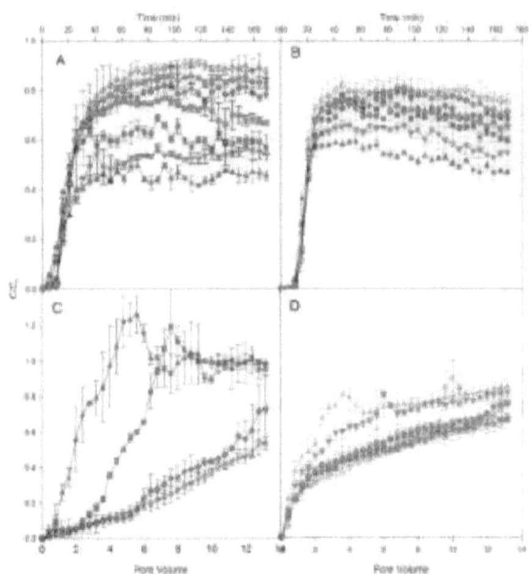

Figure S1. Breakthrough curves of *P. putida* KT2440 (Fig. S1A), *R. opacus* X9 (Fig. S1B), *P. fluorescens* LP6a (Fig. S1C), and *Sphingomonas* sp. S3 (Fig. S1D) at $E = 0$ V cm^{-1} (empty circles), $E = 0.5$ V cm^{-1} (filled hexagon), $E = 1.0$ V cm^{-1} (filled diamonds), $E = 1.5$ V cm^{-1} (filled circles), $E = 2.0$ V cm^{-1} (filled squares), $E = 2.5$ V cm^{-1} (filled downward triangles), and 3.0 (filled upward triangles) V cm^{-1}.

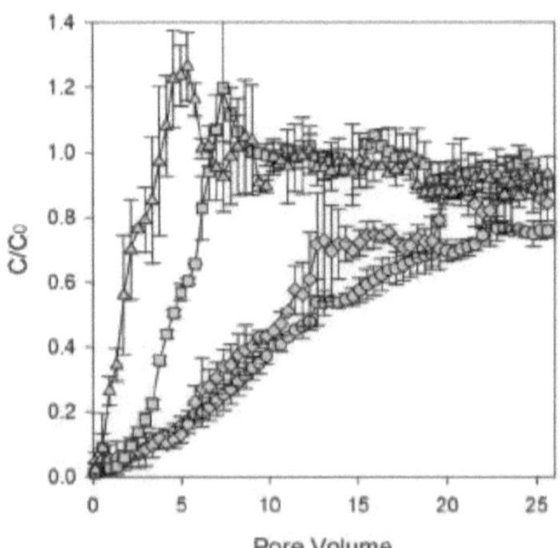

Figure S2. Breakthrough curves of *P. fluorescens* LP6a from 0 to 25 pore volumes at $E = 0$ V cm^{-1} (circles), $E = 1$ V cm^{-1} (diamonds), $E = 2$ V cm^{-1} (squares), and $E = 3$ V cm^{-1} (triangles).

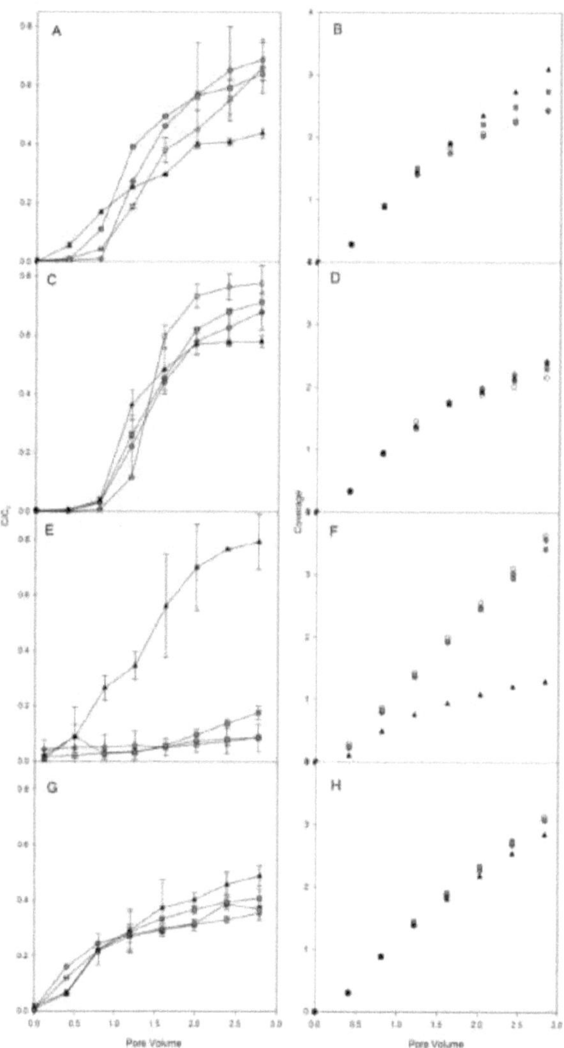

Figure S3. Breakthrough curves (left) and calculated fractions (right) of the first 2 pore volumes of four bacteria retained in percolation columns packed with glass beads in the absence (open circle) and presence (filled symbols) of DC electric fields of $E = 1.0$ V cm^{-1} (rhomboids), 2.0 V cm^{-1} (squares) and 3.0 V cm^{-1} (triangles): *P. putida* KT2440 (Figs. S3A & B), *R. opacus* X9 (Figs. S3C & D), *P. fluorescens* LP6a (Figs. S3E & F), and *Sphingomonas* sp. S3 (Figs. S3G & H). All data represent averages and standard deviations of triplicate experiments.

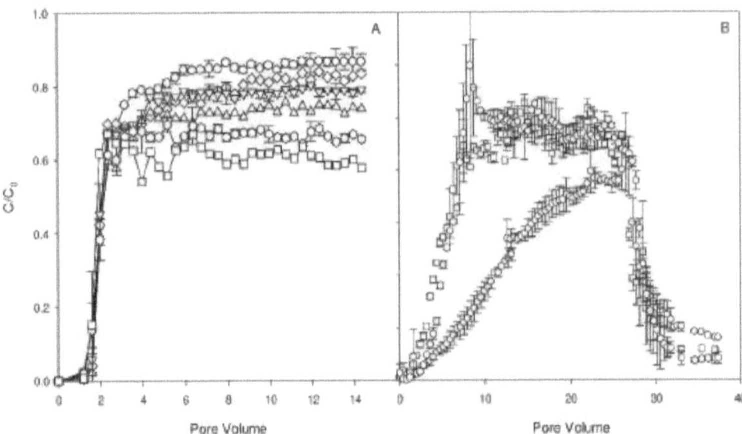

Figure S4. *P. putida* KT2440 (Fig. S4A) breakthrough curves under DC free (circles), 0.5 Vcm⁻¹ (diamonds), 1.0 V cm⁻¹ (downward triangles), 1.5 V cm⁻¹ (upward triangles), 2.0 V cm⁻¹ (hexagons), and 2.5 V cm⁻¹ (stones) with anode on top of column (reversed polarity comparing to normal experiments); and *P. fluorescens* LP6a (Fig. S4B) breakthrough curves under DC free (open circles), 2 V cm⁻¹ normal (hexagons) and reversed electric field polarity (squares).

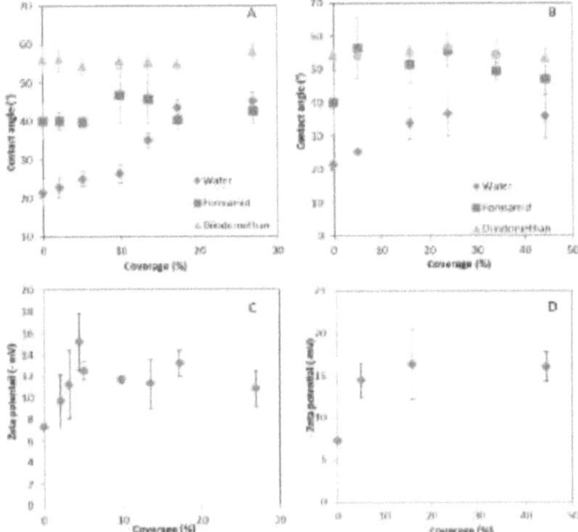

Figure S5. Effects of bacterial coverage (as estimated by eq. S16) on the zeta potential and contact angles (for water, formamide and diodomethane) of glass beads covered either by *P. putida* KT2440 (Figs. S5A & S5C) or *P. fluorescens* LP6a (Figs. S5B & S5D).

References

1. Velasco-Casal, P.; Wick, L. Y.; Ortega-Calvo, J. J., Chemoeffectors decrease the deposition of chemotactic bacteria during transport in porous media. *Environ Sci Technol* **2008**, *42*, (4), 1131-1137.

2. Martin, R. E.; Bouwer, E. J.; Hanna, L. M., Application of Clean-Bed Filtration Theory to Bacterial Deposition in Porous-Media. *Environ Sci Technol* **1992**, *26*, (5), 1053-1058.

3. Van Oss, C. J.; Good, R. J.; Chaudhury, M. K., The Role of van der Waals Forces and Hydrogen Bonds in "Hydrophobic Interactions" between Biopolymers and Low Energy Surfaces. *J Colloid Interface Sci* **1985**, *111*, (2).

4. Vilinska, A.; Rao, K. H., Surface thermodynamics and extended DLVO theory of *Leptospirillum ferrooxidans* cells' adhesion on sulfide minerals. *Miner Metall Proc* **2011**, *28*, (3), 151-158.

5. Fowkes, F. M., Attractive Forces at Interfaces. *Ind Eng Chem* **1964**, *56*, (12), 40-&.

6. Brown, D. G.; Jaffe, P. R., Effects of nonionic surfactants on the cell surface hydrophobicity and apparent hamaker constant of a *Sphingomonas sp. Environ Sci Technol* **2006**, *40*, (1), 195-201.

7. Bergström, L.; Stemme, S.; Dahlfors, T.; Arwin, H.; Ödberg, L., Spectroscopic Ellipsometry Characterisation and Estimation of the Hamaker Constant of Cellulose. *Cellulose* **1999**, *6*, (1), 1-13.

8. Johnson, W. P.; Blue, K. A.; Logan, B. E.; Arnold, R. G., Modeling Bacterial Detachment during Transport through Porous-Media as a Residence-Time-Dependent Process. *Water Resour Res* **1995**, *31*, (11), 2649-2658.

9. Jewett, D. G.; Logan, B. E.; Arnold, R. G.; Bales, R. C., Transport of *Pseudomonas fluorescens* strain P17 through quartz sand columns as a function of water content. *J Contam Hydrol* **1999**, *36*, (1-2), 73-89.

10. Rijnaarts, H. H. M.; Norde, W., Reversibility and mechanism of bacterial adhesion. *Colloids Surf B Biointerfaces* **1995**, *4*, 5-22.

11. Boks, N. P.; Norde, W.; van der Mei, H. C.; Busscher, H. J., Forces involved in bacterial adhesion to hydrophilic and hydrophobic surfaces. *Microbiol-Sgm* **2008**, *154*, 3122-3133.

12. Van Oss, C. J.; Docoslis, A.; Wu, W.; Giese, R. F., Influence of macroscopic and microscopic interactions on kinetic rate constants - I. Role of the extended DLVO theory in determining the kinetic adsorption constant of proteins in aqueous media, using von Smoluchowski's approach. *Colloids Surf B Biointerfaces* **1999**, *14*, (1-4), 99-104.

13. Vanoss, C. J.; Giese, R. F.; Costanzo, P. M., DLVO and Non-DLVO interactions in Hectorite. *Clay Clay Miner* **1990**, *38*, (2), 151-159.

14. Sharma, P. K.; Rao, K. H., Adhesion of *Paenibacillus polymyxa* on chalcopyrite and pyrite: surface thermodynamics and extended DLVO theory. *Colloids Surf B Biointerfaces* **2003**, *29*, (1), 21-38.

15. Van Oss, C. J.; Chaudhury, M. K.; Good, R. J., Interfacial Lifshitz-van der Waals and polar interactions in macroscopic systems. *Chem Rev* **1988**, *88*, (6), 927-941.

16. Meylheuc, T.; Bellon-Fontaine, M. N., *Pseudomonas fluorescens* - prodution of biosurfactants and impact of their bioadhesive behavior. **2003**.

17. Qin, J.; Sun, X.; Liu, Y.; Berthold, T.; Harms, H.; Wick, L. Y., Electrokinetic Control of Bacterial Deposition and Transport. Environ. Sci. Technol. 2015, 49 (9), 5663-5671.

4. Electrokinetic Effects on Bacterial Deposition on Planar Surfaces

4.1 Predicting Electrokinetic Effects on Bacterial Deposition by Quartz Crystal Microbalance with Dissipation Monitoring

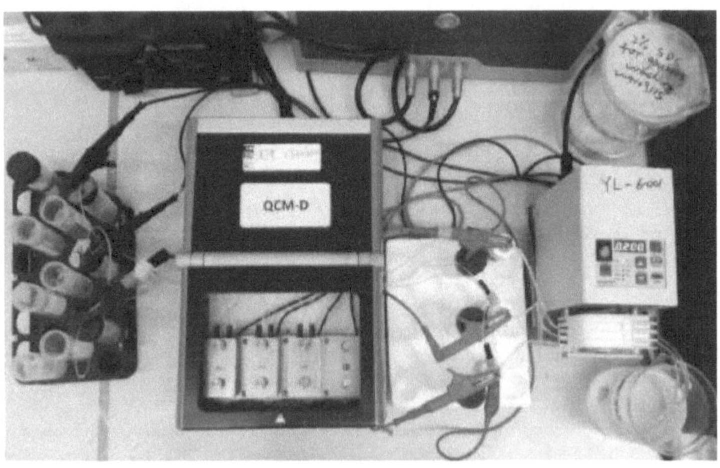

Predicting DC Electric Field Effects on Bacterial Deposition by Quartz Crystal Microbalance with Dissipation Monitoring (QCM-D)

Introduction

Microbial biofilms provide essential ecosystem services in many natural and man-made environments. While being beneficial in e.g. wastewater treatment systems or the degradation of contaminants, biofilms can also be detrimental to both human health and industrial applications. Biofouling, for instance, gives rise to unwanted corrosion of metals[1] or threatens human health by infecting medical devices[2,3] or drinking water systems[4–6]. There is hence high interest in control of bacterial deposition as the first step in the formation of biofilms. Direct current electric fields and their associated electrokinetic phenomena have been found to influence bacterial deposition[7–11]. DC electric fields evoke various electrokinetic transport processes in immersed matrices. They allow for targeted movement of bacteria and colloidal particles[19–21] in porous media also in the absence of pressure-driven hydraulic flow[10,15–17]. While electromigration and electrophoresis refer to the transport of charged molecules and particles to the electrode of opposite charge, electroosmosis (EOF) reflects the surface charge-induced movement of pore fluids usually from the anode to the cathode[18]. Due to its plug shape flow profile acting as close as a few nanometres above a surface, EOF is also thought to affect bacterial deposition by inducing shear forces (F_{EOF})[19–21]. Electrophoresis, by contrast, induces a drag force (F_{EP}) on the (negatively) charged bacteria[22–24] and hence acts in the opposite direction to F_{EOF}. A bacterium approaching a surface or being located at a distance of the secondary DLVO energy minimum will be subject to F_{EOF} and F_{EP} and the relative strength of the two forces has been proposed as a driver for observed DC-field effects on bacterial deposition[22,25–27]. Electrokinetic phenomena directly correlate to the electric field strength (E) applied, the surface properties of the matrices and (bio-)colloidal particles and the ionic strength of the electrolytes. Here we assessed the effect of DC electric fields on bacterial deposition using a quartz crystal microbalance with dissipation (QCM-D) approach. QCM-D is an accurate technique for real-time characterization of fouling and biofilm formation processes[28,29]. It is an acoustic method that reflects the amount and viscoelastic properties of an adhering mass by changes of the resonance frequency (Δf) and energy dissipation (ΔD) of an oscillating crystal coating sensor surface[30–33]. The Δf is an indicator of the bacterial mass attached to the sensor while ΔD indicates the softness of non-rigid adhesion[34,35]. Both signals vary according to the surface charge and hydrophobic properties of bacteria and the sensor surface[36–38]. Plotting ΔD versus Δf compares the induced energy dissipation per coupled unit mass: lower $\Delta f/\Delta D$ values indicate the formation of a dissipative, soft, and fluid film, while higher $\Delta f/\Delta D$ values suggest a more rigid layer of attached bacterial mass [28,39]. Hence, the Δf and ΔD changes of the QCM-D sensor allow to analyze the interactions of loosely bound layers of bacteria and their substratum given depositing cells of similar cell surface morphology[30,40]. If Δf values are supported by direct microscopy counting, QCM-D also can be used to quantify rates of bacterial attachment and, hence, to approximate time-resolved electrokinetic effects on bacterial deposition at varying environmental conditions and to compare bacterial deposition to electrokinetically induced forces (F_{EOF} and F_{EP}) acting on bacteria adjacent to a solid collector surface. Using a QCM-D approach, we here experimentally assessed the joint effects of DC electric field and ionic strength of the electrolyte on the deposition of two bacteria of differing physicochemical cell surface properties at the nanogram level[41]. QCM-D data were supported by microscopic cell counting and analyzed by a recently published theoretical approach calculating the DLVO colloidal interaction, hydraulic drag and electrokinetic forces acting on a bacterium near a collector surface in a DC electric field.

Material and Methods

Cultivation of bacteria and inoculum preparation

Pseudomonas putida KT2440 (GenBank accession No. AE015451)[42] and *Pseudomonas fluorescens* LP6a (GenBank accession No. AF525494)[43] were cultivated in minimal medium with 1.0 g L^{-1} glucose as carbon source until the early stationary phase (25 °C; rotary shaker at 150 rpm). The cultures were then centrifuged at 3000 × g and re-suspended in 10 mM (5 mmol K_2HPO_4 and 5 mmol KH_2PO_4 diluted in 1 L deionized water), 50 mM (29 mmol K_2HPO_4 and 21 mmol KH_2PO_4 diluted in 1 L DI water), and 100 mM (61 mmol K_2HPO_4 and 39 mmol KH_2PO_4 diluted in 1 L DI water) potassium phosphate buffer (PB, pH = 7) using a Vortex mixer (Vortex-Genie 2, Scientific Industries, USA) to obtain bacterial suspensions of an optical density of $OD_{600\ nm}$= 0.30.

Characterization of physicochemical properties of bacterial and sensor surfaces

The zeta-potential of bacteria (ζ_{bac}) and silica beads (ζ_s) was measured by Doppler electrophoretic light scattering analysis (Zetamaster, Malvern Instruments, Malvern, UK) with a Dip Cell Kit. The zeta potential of silica sensor surface was estimated using smashed silica beads in the appropriate PB electrolyte. Clean glass beads were smashed with a mortar and a pestle to a size of <100 μm, then heated at 200 °C in muffle furnace for 2 h, allowed to cool down to room temperature (25 °C) under sterile conditions. The contact angles (θ) of bacterial strains and the sensor in three solvents (i.e. water, formamide, and methylene iodide) were quantified using a DSA 100 drop-shape analysis system (Krüss GmbH, Hamburg, Germany) as described earlier[44,45] and are given in Table S1. Bacterial lawns were prepared by depositing bacteria from inoculated suspensions on cellulose acetate membrane filters (Millipore, 0.45 μm) and applying four droplets per filter in triplicate experiments for each solvent.

QCM-D analysis of cell deposition to the silica sensor surface

The interactions between bacterial cells and a silica surface were studied by using an E4 QCM-D unit (Q-Sense AB, Gothenburg, Sweden) with silica-coated sensor chips (QSX-303, 5 MHz, AT-cut, diameter: 14 mm, Q-Sense AB, Gothenburg, Sweden). Experiments were performed in a QCM-D system comprised of an inlet solution container, four QCM-D chambers, a buffering bottle, and a wastewater container (Fig. S1). Bacterial suspensions were pumped through QCM-D tubing by under pressure-driven flow using a digital peristaltic pump (ISM932A, Ismatec, Germany) at a fixed flow rate of 200 μL min^{-1} (flow velocity: 6×10^{-7} m s^{-1}) at 20 ± 0.2 °C (cf. Fig. S1). DC-fields fields ($E = 0.5$, 1.0, and 2.0 V. cm^{-1}) were generated by a power pack (BK Precision 9174), and connected to two Ti/Ir electrodes placed into the bacterial suspension (cathode) and buffering bottle (anode). Two copper wires (i.d.: 0.2 mm, renewed after each experiment) were connected to the Ti/Ir electrodes, and inserted cautiously from cathode until 2 mm before the entrance of QCM-D chamber, and from 2 mm after the exit of QCM-D chambers to the anode in the buffering bottle separately. PB at either 10, 50, or 100 mM was used as electrolyte and DC electric fields of either $E = 0$, 0.5, 1.0, or 2.0 V cm^{-1} were applied. Prior to the experiment, clean sterilized silica sensors were mounted in the QCM-D chamber carefully, the screws on the back of QCM-D chambers were sealed until hand-tight and then locked tightly by the snap on the base bracket. The frequency and dissipation were assured to deviate less than ± 10 % from the standard frequency and dissipation values at the overtones 1, 3, 5, 7, 9, 11, and 13 (corresponding to 5, 15, 25, 35, 45, 55, and 65 MHz), respectively. The system was stabilized by pumping ultrapure water for 20 min, followed by 40 min pumping of cell-free PB electrolyte (of equal ionic strength as for the cell suspensions) to derive the baselines, experiments were only proceeded when both baselines were stable. Bacterial suspensions of either *P. putida* KT2440 or *P. fluorescens* LP6a (in 10, 50, or 100 mM PB electrolyte each) were then pumped into the QCM-D chamber during 2 hours and the frequency and dissipation monitored simultaneously. All experiments were performed in triplicate at $E = 0$, 0.5, 1.0, and 2.0 V cm^{-1}.

After each experiment, the sensors were rinsed carefully with 1.5 mL ultrapure water in a 50 mL centrifuge tube and cells detached bacterial cells by an ultrasonic washing unit (RK255H, Bandelin electronic, Germany) for 10 min. The sensor was taken out using tweezers, disinfected in a UV chamber for 20 min and then cleaned in 50 mL 2 % SDS solution, rinsed thoroughly with ultrapure water, dried under a nitrogen stream, and sterilized for 20 min in a UV chamber following the washing protocol provided with the silica sensors.

Microscopic quantification of cells attached to the sensor

At the end of each QCM-D analysis (i.e. after 2 h) the attached bacteria cells were carefully detached from the sensor and collected in 1.5 mL water as described above. The suspension bacterial suspension was then centrifuged at $6000 \times$ g for 5 min, then 1.45 mL of the supernatant was removed, the bacterial pellet was re-suspended in the residual liquid (0.05 mL) with a Vortex mixer (Vortex-Genie 2, Scientific Industries, USA). The suspension was then injected to a Hemacytometer (Improved Neubauer 0.1 mm, Hausser Scientific, Germany) to take pictures and quantify the bacterial cell concentration by epifluorescence microscopy (Axioskop II microscope,

Zeiss, Jena, Germany) equipped with a camera (Carl Zeiss Microimaging GmbH, Germany). Images were analyzed by ImageJ software (ImageJ 1.46r, USA) for quantification of the cells. The automatic counting codes used for cell counting are listed in the supporting information, attached cell density on sensor surface (η_c) was calculated with the attached cell number on each sensor divided by the sensor surface area.

Theory

Forces acting on bacteria on a collector surface

Although the Derjaguin, Landau, Verwey, and Overbeek (DLVO) theory of colloidal interactions[46–48] does not account for surface heterogeneities, hydration effects, or hydrophobic interactions, it is a powerful predictor of bacterial deposition in solutions of high ionic strength ($I = 0.1 - 0.3$ M)[49–52]. DLVO interaction energy profiles often evolve in a characteristic fashion and depend on the physicochemical properties of the microbe, the collector surface, and the ionic strength of the aqueous medium. DLVO theory also predicts high attractive forces resulting in reversible bacterial deposition[53,54] at a so-called secondary minimum of the energy profile typically located at 5-20 nm above a collector surface. Previous work has hypothesized that other forces acting on bacteria in the secondary minimum hence may influence bacterial deposition, attachment and biofilm formation[25,44]. The net force acting on a bacterium located at the secondary minimum is estimated by a combination of the DLVO force of colloidal interaction (F_{DLVO}), the hydraulic flow shear force (F_{HF}), the electroosmotic flow shear force (F_{EOF}), and the electrophoretic drag force (F_{EP}) as described earlier[25]:

$$F_{\mathrm{net}} = F_{\mathrm{DLVO}} + F_{\mathrm{HF}} + F_{\mathrm{EOF}} + F_{\mathrm{EP}} \tag{1}$$

The calculations of the DLVO interaction force and hydraulic forces are detailed in eqs. S1-S11. It should be noted that the DLVO force is calculated at the secondary minimum distance, where the DLVO interaction controls the reversible bacterial deposition[54]. The electroosmotic shear force can be calculated by eq. 2,

$$F_{\mathrm{EOF}} = F_{\mathrm{d}}^{*} \times 6\pi\eta a V_{\mathrm{EOF}} = F_{\mathrm{d}}^{*} \times 6\pi\eta a \times [-\frac{\varepsilon_0 \varepsilon_{\mathrm{r}} \zeta_{\mathrm{s}} E}{\eta}(1 - \frac{2\mathrm{I}_1(\kappa h_{\mathrm{s}})}{\kappa a \mathrm{I}_0(\kappa h_{\mathrm{s}})})] \tag{2}$$

where F_{d}^{*} is a function of the radius a of a sphere (for simplicity we presume bacterial cells to be spheres) and the distance of the center of the sphere to the collector surface. F_{d}^{*} is estimated to be 1.7. η is the viscosity of the liquid ($\eta = 3.19$ kg m^{-1} h^{-1}), ε_{r} is the dielectric constant of water (78.5), ε_0 (8.85×10^{-12} F m^{-1}) is the vacuum permittivity, ζ_{s} is the zeta potential of the sensor surface at the experimental conditions, and E is the electric field strength applied, I_0 and I_1 are the zero- and first-order modified Bessel functions, and κ^{-1} is the thickness of the electric double layer. The electrophoretic drag force follows the Smoluchowski equation (eq. 3):

$$F_{EP} = 6\pi\eta a V_{EP} = 6\pi\eta a \times \frac{2\varepsilon_0\varepsilon_r\zeta_{bac}E}{3\eta} f(\kappa a) \tag{3}$$

Here ζ_{bac} is the zeta potential of the bacteria at given experimental conditions, the $f(\kappa a)$ values approach 1.5 in high electrolyte concentration (i.e. 50 and 100 mM), while $f(\kappa a)$ is close to 1.0 in low ionic strength (i.e. 10 mM) for the bacterial (radius $a = 0.6$ μm)[18]. The ratio of $|F_{EOF}|$ and $|F_{EP}|$ at the distance of the secondary minimum of the DLVO interaction energy above a collector surface (h_s) is detailed by eq. 4:

$$\frac{|F_{EOF}|}{|F_{EP}|} = \frac{F_d^*\zeta_s}{\frac{2}{3}\zeta_{bac}f(\kappa a)}[1 - \frac{2I_1(\kappa h_s)}{\kappa a I_0(\kappa h_s)}] \tag{4}$$

Eq.4 shows that the ratio directly depends on ζ_{bac} and ζ_s as well as κ^{-1} as the thickness of the electric double layer and, hence, is strongly influenced by the ionic strength of the electrolyte.

QCM-D analyses of bacterial deposition

QCM-D is an acoustic method that reflects the amount and viscoelastic properties of an adhering mass by changes of the resonance frequency (Δf) and energy dissipation (ΔD) of an oscillating crystal-coated sensor surface[30–33][55,56]. The mass of attachment can be described by Sauerbrey equation[57]:

$$\Delta f = \frac{-2f_0^2\Delta m}{A\sqrt{\rho_q\mu_q}} = -C_f\Delta m \tag{5}$$

where f_0 denotes the fundamental resonance frequency, A is the electrode area, ρ_q is the density of quartz ($\rho_q = 2.648$ g cm^{-3}) and μ_q is the shear modulus of quartz ($\mu_q = 2.957\times10^{10}$ N m^{-2}). The $\Delta f/\Delta D$ ratio indicates changes of energy dissipation per coupled unit mass and is an indication for the rigidity and attachment strength of bacterial adhesion[39,40,58]. Typically, bacterial adhesion leads to negative frequency shift and positive dissipation shift. Thus a less negative $\Delta f/\Delta D$ value indicates build-up of a dissipative soft and fluid film on the QCM-D sensor. More negative $\Delta f/\Delta D$ values by contrast stand for a more rigid layer.

Results

Electric field and electrolyte effects on calculated F_{net}

In order to approximate DLVO energy profiles and the electrokinetic forces acting on bacteria above a sensor surface, the sensor and bacterial physicochemical surface properties were determined in 10, 50 or 100 mM PB electrolytes. While the quartz sensor was hydrophilic (water contact angle, $\theta_w = 21°$), both strains were moderately hydrophobic ($\theta_{w,KT2440} = 70°$; $\theta_{w,LP6a} = 46°$; Table S1). The sensor surface and both bacterial strains were negatively charged in all PB electrolytes (Table 1) with more negative zeta potentials at lower ionic strengths (i.e. shifts from –21 mV (10 mM PB) to –8 mV (100 mM PB) of the sensor, -30 mV to –11 mV (strain KT2440) and

–53 mV to –36 mV (strain LP6a) (Table 1). Calculated DLVO interaction energy profiles between the bacteria and the QCM-D quartz sensor surfaces (Fig. S2) all exhibited secondary minima allowing for reversible attachment at all PB electrolyte concentrations. They were found at separation distances of 3.2 – 20.6 nm (Table S2). Corresponding attractive DLVO forces (F_{DLVO}) depended on the ionic strength of the PB and ranged from 0.15 pN (10 mM) and 3.26 pN (100 mM) for strain KT2440 and from 0.15 pN (10 mM) and 2.31 pN (100 mM) for strain LP6a, respectively (Table 1). Table 1 further summarizes the forces F_{HF}, F_{EOF}, F_{EP}, and F_{net} that we defined as the sum of the magnitudes of F_{HF}, F_{EOF}, and F_{EP} and F_{DLVO} disregarding distinct directions of the electrokinetic and DLVO forces. (eq. 1). As sensor and bacterial surfaces had negative zeta potentials (Table 1), the direction of F_{EP} was opposed to F_{EOF} and the magnitudes of F_{EP} of opposite sign to F_{EOF}. While the extent of F_{HF} was assumed independent of the experimental variations, the magnitudes of F_{EOF} and F_{EP} (expressed by $|F_{EOF}|$ and $|F_{EP}|$) increased proportionally to E (eqs. 2 & 3), decreased however at rising electrolyte concentrations. F_{net} thus depended on the electric field strength and the ionic strength of the PB electrolyte (Table 1): at any given electric field strength, higher PB concentrations increased F_{net} of both strains. At a given ionic strength, however, F_{net} of the two strains revealed dissimilar trends at increasing E: in 50 and 100 mM PB electrolyte, a rise of E from 0.5 to 2 V cm^{-1} increased F_{net} by ca. 10-20 % for strain KT2440 yet decreased F_{net} by ca. 700 % (Table 1).

Electric field and electrolyte effects on Δf and ΔD and derived cell attachment rigidity
QCM-D experiments recorded frequency and dissipation shifts at overtones 1, 3, 5, 7, 9, 11, and 13 (Fig. S3) during 120 minutes of bacterial deposition. While overtone 1 was poorly stable and overly sensitive, all other overtones showed similar trends (Figs. S4 & S5). In the following, we analyze and discuss overtone 5 as representative signal using the frequency baseline in cell-free PB electrolyte as reference to calculate the frequency and dissipation shifts (Figs. 1, S4 & S5). Figure 1 exemplifies Δf_5 and ΔD_5 shifts of both strains in 100 mM PB electrolyte at varying electric field strengths applied ($E = 0, 0.5, 1$, or 2 V cm^{-1}). Here, pumping bacteria over the sensor surface resulted in decreasing (negative) frequency shifts and increasing dissipation shifts. The extents of Δf_5 and ΔD_5, however, varied at different experimental conditions (Figs. 1A & B, S4 & S5). Generally, the rates of Δf_5 and ΔD_5 changes were higher at the beginning (0 - 15 minutes) than at the end of bacterial deposition (cf.: Figs. 1A & B for 100 mM PB and Figs. S4 & S5 for 10 and 50 mM PB) while $\Delta f_5/\Delta D_5$ ratios, as an indicator of attachment rigidity, generally exhibited highly linear correlation of Δf_5 and ΔD_5 ranges with coefficients of determination (r^2) of > 0.95 (Figs. 1C &D, Table S4). Figs. 2A-F summarize Δf_5, ΔD_5, and $\Delta f_5/\Delta D_5$ ratios at the end of the deposition experiments. While signals of strain KT2440 and LP6a differ depending on the experimental conditions chosen, our data show that observed effects were proportional to the electric field strength applied; i.e. stronger E resulted in stronger observed effects. For strain KT2440 in a 100

mM PB electrolyte, for instance, Δf_5 decreased from -18.2 Hz ($E = 0$ V cm^{-1}) to -34.5 Hz ($E = 2.0$ V cm^{-1}) while ΔD_5 increased from 2.56 ppm to 4.25 ppm (Figs. 2A & C). Such shifts resulted in clear increases of the calculated rigidity (i.e. more negative $\Delta f_5/\Delta D_5$ ratios; Fig. 2E). By contrast, Δf_5, ΔD_5 and $\Delta f_5/\Delta D_5$ ratios of strain LP6a in 100 mM PB electrolyte increased at rising electric field strengths; i.e. Δf_5 from -12.4 Hz to 3.14 Hz, ΔD_5 from 1.89 ppm to 2.34 ppm, and $\Delta f_5/\Delta D_5$ from -6.56 to -1.34 MHz (Figs. 2 B, D, F). Decreasing PB electrolyte concentrations from 100 mM to 10 mM resulted in lower shifts of Δf_5, ΔD_5 and $\Delta f_5/\Delta D_5$ in DC free controls and smaller DC-induced changes, respectively. For strain LP6a, an electric field as weak as $E = 0.5$ V cm^{-1} already resulted in distinct changes of Δf_5, ΔD_5 and $\Delta f_5/\Delta D_5$ at all PB electrolyte concentrations. By contrast, DC field effects on the trends of Δf_5, ΔD_5 and $\Delta f_5/\Delta D_5$ of strain KT2440 varied and depended on the concentration of the electrolyte. At PB concentrations of 10 and 50 mM, DC fields decreased the rigidity of attached KT2440 cells; while clearly more negative $\Delta f_5/\Delta D_5$ ratios (i.e. more rigid attachment) were found at increasing E.

Electric field and electrolyte effects on cell density of attached bacteria

The number of cells attached to the sensor surface was counted microscopically at the end of the deposition experiments and the cell density (η_c) and the surface coverage (cf. eq. S10) of cells attached to the quartz sensor surface (1.54 cm^2) were approximated. The η_c varied from 0.9×10^6 - 9.9×10^6 cells cm^{-2} (strain KT2440) and 0.7×10^6 - 3.3×10^6 cells cm^{-2} (strain LP6a) (Table 1; Figs. 2G & H). This corresponds to maximal coverages of the sensor surface (Table S3) of 1.6 % and 5.5 %, respectively. Strain LP6a excepted (where η_c at 10 mM and 50 mM were similar), the cell density increased in the order of η_c (10 mM) < η_c (50 mM) < η_c (100 mM) at all electric field strengths (Table 1). At a given PB electrolyte concentration, however, the strength of the electric fields evoked distinct η_c differences between the two bacterial strains (Table 1 and Figs. 2 E & F). Increasing E resulted in continuously decreasing η_c of strain LP6a at all electrolyte concentrations proposing that DC electric fields reduced the deposition of LP6a cells to the sensor surface even at weak E. For strain KT2440 however, the electric field decreased cell attachment to the sensor in 10 mM electrolyte, yet promoted cell attachment in 50 mM and 100 mM electrolytes (Table 1 and Figs. 2E & F). Cell density data of both strains thereby showed similar relative trends as observed by Δf_5 and ΔD_5 (Figs. 2A & B).

Discussion

Assessment of DC-induced deposition effects by QCM-D monitoring

Motivated by recent work suggesting that bacterial deposition and transport in percolation systems is influenced by electrokinetic forces[25], we here studied DC electric field effects on bacterial deposition by real-time QCM-D monitoring at varying PB electrolyte concentrations (10 - 100 mM) and electric field strengths (0 – 2 V cm^{-1}). Both variables are key drivers of the electrokinetic shear and drag forces acting on bacteria. The QCM-D signals were further compared to microscopy cell density counting and all results discussed based on approximations of the net force (F_{net}; eq. 1) acting on a bacterium at the distance of reversible attachment (i.e., at the secondary minimum of the DLVO interaction energy of bacterial adhesion, G_{DLVO}, eq. S1 and Fig. S2). Except for strain LP6a at 2 V cm^{-1} we found good correlation between the resonance frequency (Δf_5) and energy dissipation (ΔD_5) for both bacteria in all experiments (Figs 1C & D). Based on work by Gutman et al.[39] we used $\Delta f_5/\Delta D_5$ ratios as indicator of attachment rigidity[36] and cell deposition[28]. Such assumption was supported by our data that showed good correlation of $\Delta f_5/\Delta D_5$ and microscopically determined cell density (η_c) counts (Fig. 3A). Backed by both, attachment rigidity and η_c, we found that weak DC fields clearly changed the deposition patterns of strains KT2440 and LP6a as compared to DC free controls (Figs. 2E-H). Observed deposition effects were proportional to the electric field strength applied (i.e. exhibited stronger effects at higher E) yet depended on the bacterial cell surface properties and the PB electrolyte ionic strength (Fig. 2).

Prediction of DC-induced bacterial deposition effects

According to the Derjaguin, Landau, Verwey, and Overbeek (DLVO) theory[47] deposition of a bacterium to a sensor surface requires that the net kinetic energy of a bacterium is lower than DLVO interaction energy at the distance of reversible attachment[51,59]. Prediction of the DC electric field effects on bacterial deposition hence should consider additional electrokinetic forces acting on depositing cells; i.e. the electroosmotic shear and the electrophoretic drag forces as powerful tools in controlling the movement of bacteria and (bio-)colloidal particles[7,9,14]. We hence correlated DC-induced deposition effects to F_{net} shifts (Figs. 3B & S7); i.e. the attachment rigidity ($\Delta f_5/\Delta D_5$) and the cell density (η_c), to F_{net} acting on a bacterium at the secondary minimum above the sensor surface. For easier comparison all data were normalized for DC-free controls, using (($\Delta f_5/\Delta D_5$)$_{DC}$ - ($\Delta f_5/\Delta D_5$)$_{no\ DC}$) / ($\Delta f_5/\Delta D_5$)$_{no\ DC}$) i.e.: for attachment rigidity, ($\eta_{c,DC}$ - $\eta_{c\ no\ DC}$)/ $\eta_{c,no\ DC}$) for cell density, and ($F_{net,DC}$ - $F_{net,no\ DC}$)/ $F_{net,no\ DC}$) for normalized net force shifts, respectively. Doing so, we found good apparent correlation between the normalized η_c (i.e. microscopy cell counts) and QCM-D derived rigidity (Fig 3A) at all electric field strengths and buffer concentrations tested. Increasing attachment rigidity was mirrored by higher η_c, while decreasing attachment rigidity resulted in lower η_c (Fig. 3A). This highlights QCM-D as a useful approach to assess and predict the influence of DC electric fields on bacterial deposition: At $F_{net,DC} > F_{net,noDC}$ increased attachment rigidity (Fig 3B) and η_c (Fig. S7) and at $F_{net,DC} < F_{net,noDC}$ lowered attachment rigidity (Fig 3B) and η_c (Fig. S7) were observed. As F_{EOF} and F_{EP} are of opposite sign in our experimental system, their relative strength is a driver of $F_{net,DC}$ (eq.1) and, thus, of observed electrokinetic effects on bacterial deposition (Figs. 4 and S8). If $|F_{EOF}| > |F_{EP}|$, DC fields promote attachment rigidity and η_c and *vice versa*, respectively[25]. The $|F_{EOF}|$ / $|F_{EP}|$ thus was a good predictor for bacterial electrokinetic effects on cell attachment rigidity and bacterial deposition at all conditions tested. The heat maps in Figs. 4 & S8 visualize the effects of $|F_{EOF}|$ and $|F_{EP}|$ to normalized DC-induced rigidity and η_c changes. They reveal the importance of the $|F_{EP}|$ for cell deposition at given $|F_{EOF}|$ independent of the strain, electrolyte strength or the electric field applied. The high degree of convergence of rigidity and η_c changes further proposes that QCM-D is a good and fast tool for real-time analysis of electrokinetic deposition.

Relevance for environmental applications

Electrokinetic transport processes are often applied in civil and environmental engineering such as for repair and maintenance purposes or for contaminant removal. As an alternative to physical filtration, electrokinetic approaches are useful for the pre-concentration of large size molecules and nanoparticles using double layer properties of nanochannels ('electrokinetic trapping'[60]). Here we give evidence that electrokinetic forces can be applied to influence bacterial deposition to surfaces. Electrokinetic deposition approaches may promote the retention of unwanted bacteria in drinking water purification systems or, *vice versa*, may reduce bio-fouling and bio-corrosion in engineered systems. The relative strength of F_{EOF} and F_{EP} acting on bacteria at a distance of the secondary DLVO minimum above a surface was found to be a good predictor for electrokinetic effects on cell deposition. According to eq. 4 the $|F_{EOF}|/|F_{EP}|$ ratio is influenced by the electric field strength, the ionic strength of the electrolyte, the zeta potential of the bacteria and the collector surfaces, and the thickness of the electric double layer. QCM-D allows for fast, real-time and accurate high throughput monitoring of bacterial deposition by easily changing the drivers of the $|F_{EOF}|$ / $|F_{EP}|$ ratio. It hence can be used to back the predicted electrokinetic effects on bacterial deposition prior to further environmental and biotechnological applications (e.g. retention of unwanted bacteria in drinking water purification or the prevention of biofilm induced corrosion). Knowledge on DC-effects also allows to manage electrokinetic bacterial dispersal in subsurface porous media and e.g. to change microbial community structures and functions and to promote contaminant biodegradation in disturbed ecosystems[61,62]. In parallel, electrokinetic effects may simultaneously improve the transport of nutrients by electromigration or change the interactions of contaminants with sorbents[63,64] and thereby improve their bioavailability and biodegradation during engineered clean-up of contaminated soil or waters.

Acknowledgments. This work has been performed in the frame of the Helmholtz Alberta Initiative and contributes to the research program topic CITE of the Helmholtz Association. We acknowledge financial support by the China Scholarship Council (CSC) and the German Academic Exchange Service (DAAD). The authors wish to thank Luis Rosa for helpful discussions as well as Jana Reichenbach, Rita Remer and Birgit Würz for skilled technical help.

References

(1) Iverson, W. P. Microbial Corrosion of Metals. In *Advances in Applied Microbiology*; Laskin, A. I., Ed.; Academic Press, 1987; Vol. 32, pp 1–36. https://doi.org/10.1016/S0065-2164(08)70077-7.

(2) Russotto, V.; Cortegiani, A.; Raineri, S. M.; Giarratano, A. Bacterial Contamination of Inanimate Surfaces and Equipment in the Intensive Care Unit. *J. Intensive Care* **2015**, *3* (1), 54. https://doi.org/10.1186/s40560-015-0120-5.

(3) Fafliora, E.; Bampalis, V. G.; Lazarou, N.; Mantzouranis, G.; Anastassiou, E. D.; Spiliopoulou, I.; Christofidou, M. Bacterial Contamination of Medical Devices in a Greek Emergency Department: Impact of Physicians' Cleaning Habits. *Am. J. Infect. Control* **2014**, *42* (7), 807–809. https://doi.org/10.1016/j.ajic.2014.03.017.

(4) Douterelo, I.; Jackson, M.; Solomon, C.; Boxall, J. Microbial Analysis of in Situ Biofilm Formation in Drinking Water Distribution Systems: Implications for Monitoring and Control of Drinking Water Quality. *Appl. Microbiol. Biotechnol.* **2016**, *100* (7), 3301–3311. https://doi.org/10.1007/s00253-015-7155-3.

(5) Lamka, K. G.; Lechevallier, M. W.; Seidler, R. J. Bacterial Contamination of Drinking Water Supplies in a Modern Rural Neighborhoodt. *APPL Env. MICROBIOL* **1980**, *39*, 5.

(6) *Guidelines for Drinking-Water Quality*, 4th ed.; World Health Organization, Ed.; World Health Organization: Geneva, 2011.

(7) Masliyah, J. H.; Bhattacharjee, S. *Electrokinetic and Colloid Transport Phenomena*; John Wiley & Sons: New Jersey, 2006.

(8) *Interfacial Electrokinetics and Electrophoresis*; Delgado, Á. V., Ed.; Surfactant science series; Marcel Dekker, Inc: New York, 2002.

(9) Shi, L.; Müller, S.; Harms, H.; Wick, L. Y. Factors Influencing the Electrokinetic Dispersion of PAH-Degrading Bacteria in a Laboratory Model Aquifer. *Appl. Microbiol. Biotechnol.* **2008**, *80* (3), 507–515. https://doi.org/10.1007/s00253-008-1577-0.

(10) Shi, L.; Müller, S.; Harms, H.; Wick, L. Y. Effect of Electrokinetic Transport on the Vulnerability of PAH-Degrading Bacteria in a Model Aquifer. *Environ. Geochem. Health* **2008**, *30* (2), 177–182. https://doi.org/10.1007/s10653-008-9146-0.

(11) Wick, L. Y.; Shi, L.; Harms, H. Electro-Bioremediation of Hydrophobic Organic Soil-Contaminants: A Review of Fundamental Interactions. *Electrochimica Acta* **2007**, *52* (10), 3441–3448. https://doi.org/10.1016/j.electacta.2006.03.117.

(12) Hu, Y.; Werner, C.; Li, D. Electrokinetic Transport through Rough Microchannels. *Anal. Chem.* **2003**, *75* (21), 5747–5758. https://doi.org/10.1021/ac0347157.

(13) Kuo, C. C.; Papadopoulos, K. D. Electrokinetic: Movement of Settled Spherical Particles in Fine Capillaries. *Environ. Sci. Technol.* **1996**, *30* (4), 1176–1179. https://doi.org/10.1021/es950413d.

(14) Hunter, R. J. *Zeta Potential in Colloid Science: Principles and Applications*, 3. print.; Colloid science; Academic Pr: London, 1988.

(15) Wick, L. Y.; Mattle, P. A.; Wattiau, P.; Harms, H. Electrokinetic Transport of PAH-Degrading Bacteria in Model Aquifers and Soil. *Environ. Sci. Technol.* **2004**, *38* (17), 4596–4602. https://doi.org/10.1021/es0354420.

(16) Secord, E. L.; Kottara, A.; Van Cappellen, P.; Lima, A. T. Inoculating Bacteria into Polycyclic Aromatic Hydrocarbon-Contaminated Oil Sands Soil by Means of Electrokinetics. *Water. Air. Soil Pollut.* **2016**, *227* (8), 288.

(17) Haber, S. Deep Electrophoretic Penetration and Deposition of Ceramic Particles inside Impermeable Porous Substrates. *J. Colloid Interface Sci.* **1996**, *179* (2), 380–390.

(18) Elimelech, M. Particle Deposition on Ideal Collectors from Dilute Flowing Suspensions: Mathematical Formulation, Numerical Solution, and Simulations. *Sep. Technol.* **1994**, *4* (4), 186–212. https://doi.org/10.1016/0956-9618(94)80024-3.

(19) Ross, D.; Johnson, T. J.; Locascio, L. E. Imaging of Electroosmotic Flow in Plastic Microchannels. *Anal. Chem.* **2001**, *73* (11), 2509–2515. https://doi.org/10.1021/ac001509f.

(20) Herr, A. E.; Molho, J. I.; Santiago, J. G.; Mungal, M. G.; Kenny, T. W.; Garguilo, M. G. Electroosmotic Capillary Flow with Nonuniform Zeta Potential. *Anal. Chem.* **2000**, *72* (5), 1053–1057. https://doi.org/10.1021/ac990489i.

(21)   Tallarek, U.; Rapp, E.; Scheenen, T.; Bayer, E.; Van As, H. Electroosmotic and Pressure-Driven Flow in Open and Packed Capillaries: Velocity Distributions and Fluid Dispersion. *Anal. Chem.* **2000**, *72* (10), 2292–2301. https://doi.org/10.1021/ac991303i.

(22)   Lee, Y.-F.; Huang, Y.-F.; Tsai, S.-C.; Lai, H.-Y.; Lee, E. Electrophoretic and Electroosmotic Motion of a Charged Spherical Particle within a Cylindrical Pore Filled with Debye–Bueche–Brinkman Polymeric Solution. *Langmuir* **2016**, *32* (49), 13106–13115. https://doi.org/10.1021/acs.langmuir.6b02795.

(23)   Lee, T. C.; Keh, H. J. Electrophoretic Motion of a Charged Particle in a Charged Cavity. *Eur. J. Mech. - BFluids* **2014**, *48*, 183–192. https://doi.org/10.1016/j.euromechflu.2014.06.004.

(24)   Liu, H.; Bau, H. H.; Hu, H. H. Electrophoresis of Concentrically and Eccentrically Positioned Cylindrical Particles in a Long Tube. *Langmuir* **2004**, *20* (7), 2628–2639. https://doi.org/10.1021/la035849i.

(25)   Shan, Y.; Harms, H.; Wick, L. Y. Electric Field Effects on Bacterial Deposition and Transport in Porous Media. *Environ. Sci. Technol.* **2018**, *52* (24), 14294–14301. https://doi.org/10.1021/acs.est.8b03648.

(26)   Locke, B. R. Electrophoretic Transport in Porous Media: A Volume-Averaging Approach. *Ind. Eng. Chem. Res.* **1998**, *37* (2), 615–625.

(27)   Pennathur, S.; Santiago, J. G. Electrokinetic Transport in Nanochannels. 1. Theory. *Anal. Chem.* **2005**, *77* (21), 6772–6781. https://doi.org/10.1021/ac050835y.

(28)   Schofield, A. L.; Rudd, T. R.; Martin, David. S.; Fernig, D. G.; Edwards, C. Real-Time Monitoring of the Development and Stability of Biofilms of Streptococcus Mutans Using the Quartz Crystal Microbalance with Dissipation Monitoring. *Biosens. Bioelectron.* **2007**, *23* (3), 407–413. https://doi.org/10.1016/j.bios.2007.05.001.

(29)   Herzberg, M.; Sweity, A.; Brami, M.; Kaufman, Y.; Freger, V.; Oron, G.; Belfer, S.; Kasher, R. Surface Properties and Reduced Biofouling of Graft-Copolymers That Possess Oppositely Charged Groups. *Biomacromolecules* **2011**, *12* (4), 1169–1177. https://doi.org/10.1021/bm101470y.

(30)   Olsson, A. L. J.; van der Mei, H. C.; Busscher, H. J.; Sharma, P. K. Acoustic Sensing of the Bacterium–Substratum Interface Using QCM-D and the Influence of Extracellular Polymeric Substances. *J. Colloid Interface Sci.* **2011**, *357* (1), 135–138. https://doi.org/10.1016/j.jcis.2011.01.035.

(31)   Strauss, J.; Liu, Y.; Camesano, T. A. Bacterial Adhesion to Protein-Coated Surfaces: An AFM and QCM-D Study. *JOM J. Miner. Met. Mater. Soc.* **2009**, *61* (9), 71–74.

(32)   Jiang, D.; Li, B.; Jia, W.; Lei, Y. Effect of Inoculum Types on Bacterial Adhesion and Power Production in Microbial Fuel Cells. *Appl. Biochem. Biotechnol.* **2010**, *160* (1), 182–196. https://doi.org/10.1007/s12010-009-8541-z.

(33)   Camesano, T. A.; Liu, Y.; Datta, M. Measuring Bacterial Adhesion at Environmental Interfaces with Single-Cell and Single-Molecule Techniques. *Adv. Water Resour.* **2007**, *30* (6–7), 1470–1491. https://doi.org/10.1016/j.advwatres.2006.05.023.

(34)   Fredriksson, C.; Kihlman, S.; Rodahl, M.; Kasemo, B. The Piezoelectric Quartz Crystal Mass and Dissipation Sensor: A Means of Studying Cell Adhesion. *Langmuir* **1998**, *14* (2), 248–251. https://doi.org/10.1021/la971005l.

(35)   Feiler, A. A.; Sahlholm, A.; Sandberg, T.; Caldwell, K. D. Adsorption and Viscoelastic Properties of Fractionated Mucin (BSM) and Bovine Serum Albumin (BSA) Studied with Quartz Crystal Microbalance (QCM-D). *J. Colloid Interface Sci.* **2007**, *315* (2), 475–481. https://doi.org/10.1016/j.jcis.2007.07.029.

(36)   Olsson, A. L. J.; van der Mei, H. C.; Busscher, H. J.; Sharma, P. K. Novel Analysis of Bacterium−Substratum Bond Maturation Measured Using a Quartz Crystal Microbalance. *Langmuir* **2010**, *26* (13), 11113–11117. https://doi.org/10.1021/la100896a.

(37)   Qi, M.; Gong, X.; Wu, B.; Zhang, G. Landing Dynamics of Swimming Bacteria on a Polymeric Surface: Effect of Surface Properties. *Langmuir* **2017**, *33* (14), 3525–3533. https://doi.org/10.1021/acs.langmuir.7b00439.

(38)   Song, L.; Sjollema, J.; Sharma, P. K.; Kaper, H. J.; van der Mei, H. C.; Busscher, H. J. Nanoscopic Vibrations of Bacteria with Different Cell-Wall Properties Adhering to Surfaces under Flow and Static Conditions. *ACS Nano* **2014**, *8* (8), 8457–8467. https://doi.org/10.1021/nn5030253.

(39)   Gutman, J.; Walker, S. L.; Freger, V.; Herzberg, M. Bacterial Attachment and Viscoelasticity: Physicochemical and Motility Effects Analyzed Using Quartz Crystal Microbalance with Dissipation (QCM-D). *Environ. Sci. Technol.* **2013**, *47* (1), 398–404. https://doi.org/10.1021/es303394w.

(40)   Marcus, I. M.; Herzberg, M.; Walker, S. L.; Freger, V. Pseudomonas Aeruginosa Attachment on QCM-D Sensors: The Role of Cell and Surface Hydrophobicities. *Langmuir* **2012**, *28* (15), 6396–6402. https://doi.org/10.1021/la300333c.

(41)   Tanner, F. W. *Shape and Size of Bacterial Cells: Bacteriology*; John Wiley and Sons, Inc., New York, 1948.

(42)   Nelson, K. E.; Weinel, C.; Paulsen, I. T.; Dodson, R. J.; Hilbert, H.; Martins dos Santos, V. A. P.; Fouts, D. E.; Gill, S. R.; Pop, M.; Holmes, M. Complete Genome Sequence and Comparative Analysis of the Metabolically Versatile Pseudomonas Putida KT2440. *Environ. Microbiol.* **2002**, *4* (12), 799–808.

(43)   Foght, J. M.; Westlake, D. W. Transposon and Spontaneous Deletion Mutants of Plasmid-Borne Genes Encoding Polycyclic Aromatic Hydrocarbon Degradation by a Strain of Pseudomonas Fluorescens. *Biodegradation* **1996**, *7* (4), 353–366.

(44)   Qin, J.; Sun, X.; Liu, Y.; Berthold, T.; Harms, H.; Wick, L. Y. Electrokinetic Control of Bacterial Deposition and Transport. *Environ. Sci. Technol.* **2015**, *49* (9), 5663–5671. https://doi.org/10.1021/es506245y.

(45)   Ghanem, N.; Kiesel, B.; Kallies, R.; Harms, H.; Chatzinotas, A.; Wick, L. Y. Marine Phages As Tracers: Effects of Size, Morphology, and Physico–Chemical Surface Properties on Transport in a Porous Medium. *Environ. Sci. Technol.* **2016**, *50* (23), 12816–12824. https://doi.org/10.1021/acs.est.6b04236.

(46)   Hermansson, M. The DLVO Theory in Microbial Adhesion. *Colloids Surf. B Biointerfaces* **1999**, *14* (1), 105–119.

(47)   *Particle Deposition and Aggregation: Measurement, Modelling and Simulation*; Elimelech, M., Ed.; Colloid and surface engineering series; Butterworth-Heinemann: Oxford, 1998.

(48)   Probstein, R. F. *Physicochemical Hydrodynamics: An Introduction*; John Wiley & Sons, 2005.

(49)   Redman, J. A.; Walker, S. L.; Elimelech, M. Bacterial Adhesion and Transport in Porous Media: Role of the Secondary Energy Minimum. *Environ. Sci. Technol.* **2004**, *38* (6), 1777–1785. https://doi.org/10.1021/es0348871.

(50)   Song, L.; Johnson, P. R.; Elimelech, M. Kinetics of Colloid Deposition onto Heterogeneously Charged Surfaces in Porous Media. *Environ. Sci. Technol.* **1994**, *28* (6), 1164–1171.

(51)   Simoni, S. F.; Bosma, T. N. P.; Harms, H.; Zehnder, A. J. B. Bivalent Cations Increase Both the Subpopulation of Adhering Bacteria and Their Adhesion Efficiency in Sand Columns. *Environ. Sci. Technol.* **2000**, *34* (6), 1011–1017. https://doi.org/10.1021/es990476m.

(52)   Rijnaarts, H. H. *Interactions between Bacteria and Solid Surfaces in Relation to Bacterial Transport in Porous Media*.

(53)   Rijnaarts, H. H. M.; Norde, W.; Bouwer, E. J.; Lyklema, J.; Zehnder, A. J. B. Reversibility and Mechanism of Bacterial Adhesion. *Colloids Surf. B Biointerfaces* **1995**, *4* (1), 5–22. https://doi.org/10.1016/0927-7765(94)01146-V.

(54)   Tufenkji, N.; Elimelech, M. Breakdown of Colloid Filtration Theory: Role of the Secondary Energy Minimum and Surface Charge Heterogeneities. *Langmuir* **2005**, *21* (3), 841–852. https://doi.org/10.1021/la048102g.

(55)   Ward, M. D.; Buttry, D. A. In Situ Interfacial Mass Detection with Piezoelectric Transducers. *Science* **1990**, *249* (4972), 1000–1007.

(56)   Reviakine, I.; Johannsmann, D.; Richter, R. P. Hearing What You Cannot See and Visualizing What You Hear: Interpreting Quartz Crystal Microbalance Data from Solvated Interfaces. *Anal. Chem.* **2011**, *83* (23), 8838–8848. https://doi.org/10.1021/ac201778h.

(57)   Sauerbrey, G. Verwendung von Schwingquarzen zur Wägung dünner Schichten und zur Mikrowägung. *Z. Für Phys.* **1959**, *155* (2), 206–222. https://doi.org/10.1007/BF01337937.

(58)   Kao, W.-L.; Chang, H.-Y.; Lin, K.-Y.; Lee, Y.-W.; Shyue, J.-J. Effect of Surface Potential on the Adhesion Behavior of NIH3T3 Cells Revealed by Quartz Crystal Microbalance with Dissipation Monitoring (QCM-D). *J. Phys. Chem. C* **2017**, *121* (1), 533–541. https://doi.org/10.1021/acs.jpcc.6b11217.

(59) Simoni, S. F.; Harms, H.; Bosma, T. N. P.; Zehnder, A. J. B. Population Heterogeneity Affects Transport of Bacteria through Sand Columns at Low Flow Rates. *Environ. Sci. Technol.* **1998**, *32* (14), 2100–2105. https://doi.org/10.1021/es970936g.

(60) de la Guardia, M.; Garrigues, S. *Handbook of Green Analytical Chemistry*; John Wiley & Sons, 2012.

(61) König, S.; Worrich, A.; Banitz, T.; Centler, F.; Harms, H.; Kästner, M.; Miltner, A.; Wick, L. Y.; Thullner, M.; Frank, K. Spatiotemporal Disturbance Characteristics Determine Functional Stability and Collapse Risk of Simulated Microbial Ecosystems. *Sci. Rep.* **2018**, *8* (1), 9488.

(62) König, S.; Worrich, A.; Banitz, T.; Harms, H.; Kästner, M.; Miltner, A.; Wick, L. Y.; Frank, K.; Thullner, M.; Centler, F. Functional Resistance to Recurrent Spatially Heterogeneous Disturbances Is Facilitated by Increased Activity of Surviving Bacteria in a Virtual Ecosystem. *Front. Microbiol.* **2018**, *9*, 734.

(63) Shan, Y.; Qin, J.; Harms, H.; Wick, L. Y. Electrokinetic Effects on the Interaction of Phenanthrene with Geo-Sorbents. *Chemosphere* **2019**, 125161.

(64) Qin, J.; Moustafa, A.; Harms, H.; El-Din, M. G.; Wick, L. Y. The Power of Power: Electrokinetic Control of PAH Interactions with Exfoliated Graphite. *J. Hazard. Mater.* **2015**, *288*, 25–33. https://doi.org/10.1016/j.jhazmat.2015.02.008.

Figure Legends

Figure 1. Time dependent frequency (Δf_5) and dissipation shifts (ΔD_5) of *P. putida* KT2440 (Fig. 2A) and *P. fluorescens* LP6a (Fig. 2B) at overtone 5 in 100 mM PB electrolyte and electric field strengths of $E = 0$ V cm^{-1} (empty squares), $E = 0.5$ V cm^{-1} (light gray triangles), $E = 1.0$ V cm^{-1} (dark gray circles), and $E = 2.0$ V cm^{-1} (black diamonds). Error bars denote the standard deviation of the mean ($n = 3$). Data above and below the dashed line refer to Δf_5 (left y-axis) and to ΔD_5 (right y-axis), respectively. Panels C & D correlate time dependent ΔD_5 and Δf_5 of *P. putida* KT2440 and *P. fluorescens* LP6a.

Figure 2. Effect of the electric field strength on the frequency shift (Δf_5; Figs. 2A & 2B), the dissipation shift (ΔD_5; Figs 2C & D), the rigidity of bacterial attachment ($\Delta f_5/\Delta D_5$, Figs. 2E & F), and cell density on the sensor surface (Figs. 2G & H). Data reflect bacterial deposition after two hours (cf. Fig. S2) at overtone 5 in 10 mM (light gray), 50 mM (dark gray) and 100 mM (black) PB. Figs. 2A, C, E & G reflect *P. putida* KT2440 and Figs. 2B, D, F & H reflect *P. fluorescens* LP6a.

Figure 3. Correlation of normalized changes of DC-induced cell density and rigidity of cell attachment (Fig. 3A) and DC-induced net force ($F_{net,DC}$, cf. eq 1) and rigidity of cell attachment (Fig. 3B), respectively. All plots reflect data after two hours of deposition of *P. putida* KT2440 (squares) and *P. fluorescens* LP6a (diamonds) exposed to PB electrolyte of either 10 mM (light gray), 50 mM (dark gray), and 100 mM (black) and DC electric field strengths of $E = 0$, 0.5, 1.0, or 2.0 V cm^{-1} (cf. digits at the symbols).

Figure 4. Calculated effects of the electroosmotic shear $|F_{EOF}|$ and the electrophoretic drag force $|F_{EP}|$ on DC-induced normalized changes of the rigidity of attachment after two hours of deposition of *P. putida* KT2440 (squares) and *P. fluorescens* LP6a (diamonds). Experiments were performed in PB electrolyte of 10 mM (light gray), 50 mM (dark gray), and 100 mM (black), and DC electric field strengths of $E = 0$, 0.5, 1.0, or 2.0 V cm^{-1} (cf. digits at the symbols). Data points above ($|F_{EP}| > |F_{EOF}|$) and below ($|F_{EP}| < |F_{EOF}|$) the dashed line refer to decreased and increased rigidity, respectively as compared to DC-free controls.

P. putida KT2440
P. fluorescens LP6a
A
B
C
D
2 V cm⁻¹
0.5 V cm⁻¹
1 V cm⁻¹
0 V cm⁻¹
0 V cm⁻¹
0.5 V cm⁻¹
2 V cm⁻¹
1 V cm⁻¹
Time (min)
Time (min)

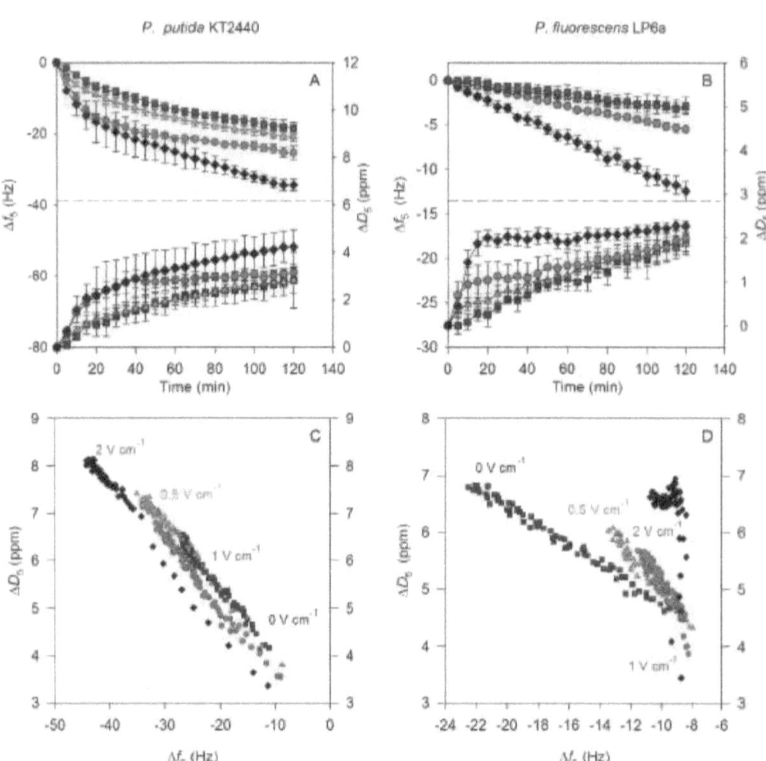

Figure 1

P. putida KT2440
P. fluorescens LP6a
A
10 mM
50 mM
100 mM
Frequency shift (Hz)
B
10 mM
50 mM
100 mM
C
100 mM
50 mM
10 mM
Dissipation shift (ppm)
D
100 mM
50 mM
10 mM
E
10 mM
50 mM
Increasing deposition
100 mM
Rigidity (*10^6 Hz)
F
10 mM
50 mM
100 mM
Decreasing deposition
G
Increasing deposition
100 mM
50 mM
10 mM
Cell density (*10^6 cells/cm^2)
H
100 mM
50 mM
10 mM
Decreasing deposition
Electric field strength (V cm^-1)
Electric field strength (V cm^-1)

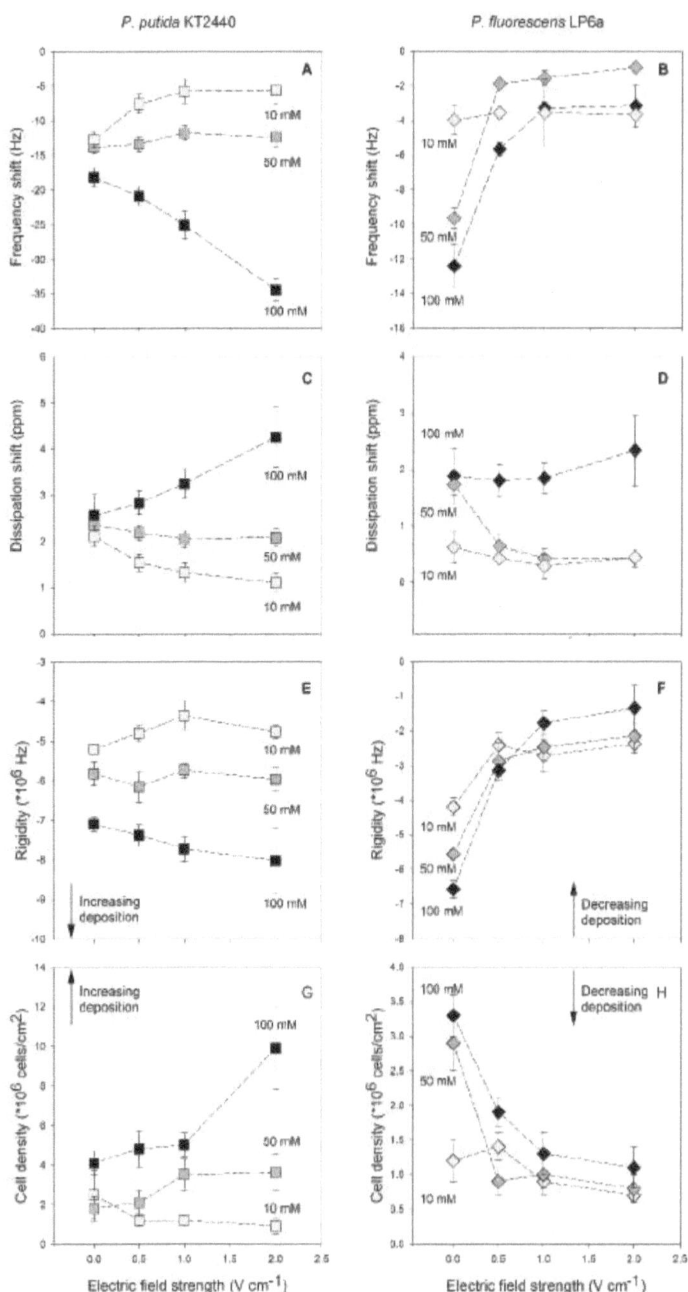

Figure 2

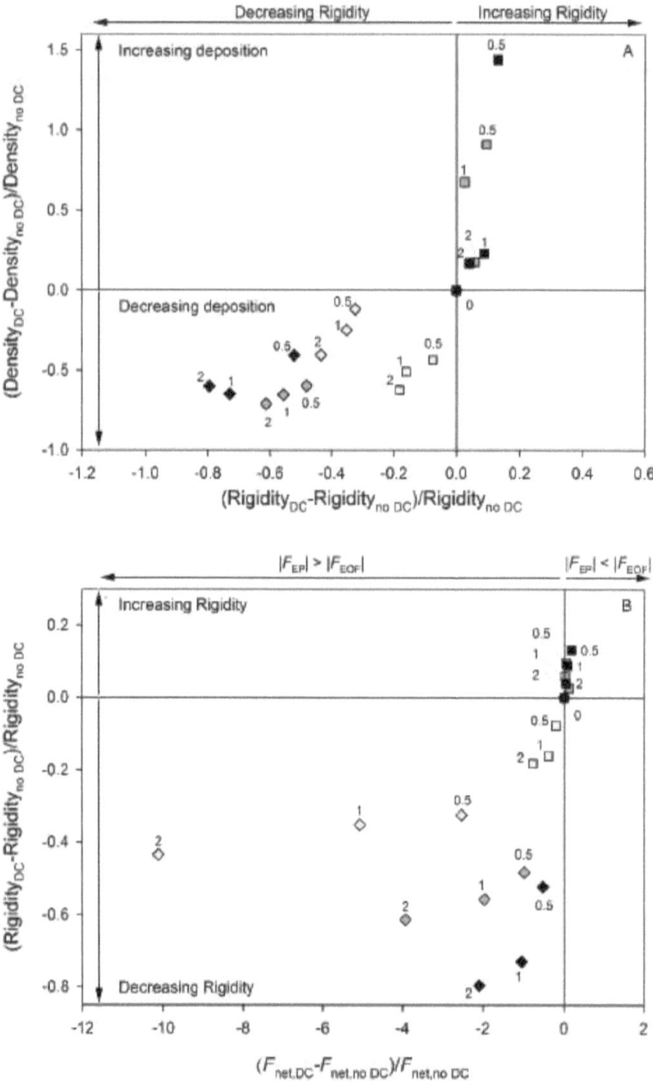

Figure 3

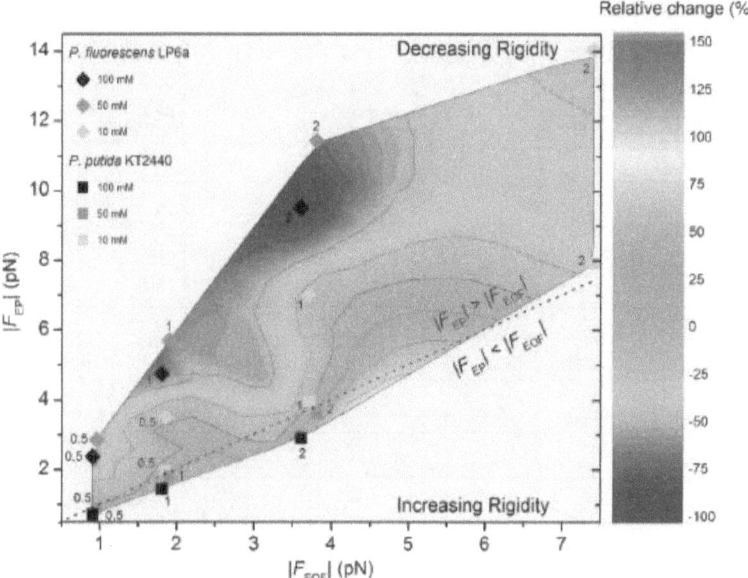

Relative change (%)
P. fluorescens LP6a
100 mM
50 mM
10 mM
P. putida KT2440
100 mM
50 mM
10 mM
Decreasing Rigidity
Increasing Rigidity
$|F_{EP}| > |F_{EOF}|$
$|F_{EP}| < |F_{EOF}|$
$|F_{EP}|$ (pN)
$|F_{EOF}|$ (pN)
14
12
10
8
6
4
2
150
125
100
75
50
25
0
-25
-50
-75
-100

Figure 4

Table 1. Overview of cell counts, zeta potential and the calculated forces acting on a bacterium (*P. putida* KT2440 and *P. fluorescens* LP6a) at the distance of the secondary minimum in presence and absence of a DC electric and different electrolyte strengths.

		P. putida KT2440			*P. fluorescens* LP6a		
		10 mM	50 mM	100 mM	10 mM	50 mM	100 mM
DLVO force (*p*N) [a]	F_{DLVO}	0.15	1.45	3.26	0.15	1.43	2.31
Hydraulic shear force (pN) [b]	F_{HF}	0.50	0.50	0.50	0.50	0.50	0.50
Electroosmotic shear force per V cm^{-1}(pN)	F_{EOF}	3.70	1.90	1.80	3.70	1.90	1.80
Electrophoretic drag force per V cm^{-1} (*p*N)	F_{EP}	-3.95	-1.80	-1.45	-6.99	-5.69	-4.74
Net force (pN) [c]							
$E = 0$ V cm^{-1}	$F_{\mathrm{net,ND}}$	0.65	1.95	3.76	0.65	1.93	2.81
$E = 0.5$ V cm^{-1}	$F_{\mathrm{net,0.5V\,cm^{-1}}}$	0.53	2.00	3.94	-1.00	0.03	1.34
$E = 1$ V cm^{-1}	$F_{\mathrm{net,1V\,cm^{-1}}}$	0.40	2.05	4.11	-2.65	-1.86	-0.13
$E = 2$ V cm^{-1}	$F_{\mathrm{net,2V\,cm^{-1}}}$	0.15	2.15	4.46	-5.95	-5.65	-3.07
Cell density (10^6 cells cm^{-2}) [d]	η_c						
$E = 0.0$ V cm^{-1}	$\eta_{c\ \mathrm{no\ DC}}$	2.5 ± 0.2	1.8 ± 0.6	4.1 ± 0.6	1.4 ± 0.2	2.9 ± 0.4	3.3 ± 0.3
$E = 0.5$ V cm^{-1}	$\eta_{c\ \mathrm{0.5V\,cm^{-1}}}$	1.2 ± 0.3	2.1 ± 0.6	4.8 ± 0.9	1.2 ± 0.3	0.9 ± 0.2	1.9 ± 0.2
$E = 1$ V cm^{-1}	$\eta_{c\ \mathrm{1V\,cm^{-1}}}$	1.2 ± 0.2	3.5 ± 0.8	5.0 ± 0.6	0.9 ± 0.1	1.0 ± 0.3	1.3 ± 0.3
$E = 2$ V cm^{-1}	$\eta_{c\ \mathrm{2V\,cm^{-1}}}$	0.9 ± 0.4	3.6 ± 0.9	9.9 ± 2.1	0.7 ± 0.1	0.8 ± 0.2	1.1 ± 0.3
Zeta potential (-mV)							
Bacteria	ζ_{bac}	-30 ± 1	-14 ± 2	-11 ± 1	-53 ± 2	-43 ± 2	-36 ± 3
		Sensor surface					
Silica [e]	ζ_{s}	-21 ± 2	-12 ± 1	-8 ± 1			

[a] For calculation cf. eq. S10; [b] F_{HF} calculated for flow velocity of 6×10^{-7} m s^{-1} (cf. eq. S11); [c] cf. eq. 1; [d] Microscopically determined cell counts after 2 h; [e] Silica sensor surface.

4.2 Supporting Information

Predicting DC Electric Fields Effects on Bacterial Deposition by Quartz Crystal Microbalance with Dissipation Monitoring (QCM-D)

1   *UFZ - Helmholtz Centre for Environmental Research, Department of Environmental Microbiology, 04318 Leipzig, Germany*

2   *University of Alberta, Department of Civil and Environmental Engineering, 3-133 Markin/CNRL Natural Resources Engineering Facility, Edmonton, AB, T6G 2W2, Canada*

Calculation of DLVO interaction force between bacteria and a solid surface (F_{DLVO})

According to the DLVO theory, the DLVO interaction energy of bacterial adhesion (G_{DLVO}) the electrostatic repulsion (G_{EDL}), and the Lifshitz-van der Waals (G_{LW}) energy (eq. S1)[1]:

$$G_{DLVO} = G_{EDL} + G_{LW} \tag{S1}$$

The surface Gibbs free energies of bacteria γ_b and the glass surface γ_s (mJ m^{-2}) were calculated based on measured contact angles (θ) of microbial lawns, membrane filters and glass surfaces using water, formamide and methylene iodide as liquids using the Young equation according to eq. S2:

$$Cos(\theta) = -1 + 2\frac{\sqrt{\gamma_b^{LW}\gamma_l^{LW}}}{\gamma_l^{total}} + 2\frac{\sqrt{\gamma_b^+\gamma_l^-}}{\gamma_l^{total}} + 2\frac{\sqrt{\gamma_b^-\gamma_l^+}}{\gamma_l^{total}} \tag{S2}$$

The total surface Gibbs free energies (γ^{total}) thereby were separated in a Lifshitz-van der Waals (γ^{LW}) and an acid-base component (γ^{AB}) (eq. S11) with γ^+ and γ^- as the electron acceptor and the electron donor components of the acid-base surface energy (eqs. S3 & 4).

$$\gamma^{total} = \gamma^{AB} + \gamma^{LW} \tag{S3}$$

$$\gamma_i^{AB} = 2\sqrt{\gamma_i^+\gamma_i^-} \tag{S4}$$

Using literature data[2] of γ, γ^{LW}, γ^+, γ^- values for water, formamide and methyleneiodide, the parameters γ_b, γ_b^{LW}, γ_b^+, γ_b^- of bacteria were calculated as proposed by van Oss et al[3], and the data from literature were taken for assessing the free energy of the glass surface.

Hamaker constant[4] can be described by eq. S5

$$A_{132} = (\sqrt{A_{11}} - \sqrt{A_{33}})(\sqrt{A_{22}} - \sqrt{A_{33}}) \tag{S5}$$

Here, A_{ii} denotes the individual Hamaker constant of bacteria (A_{11}), glass (A_{22},) and water (A_{33}), respectively. A_{33} was taken from literature[5] whereas A_{11} and A_{22} were obtained by eq. S6

$$A_{ii} = 6\pi l_0^2 \gamma_i^{LW} \tag{S6}$$

According to Fowkes[6], the value of $6\pi l_0^2$ equals 1.44×10^{-18} m^2, with l_0 being the minimum distance between the outermost cell surface and the glass bead (0.157 nm)[7].

The electrostatic repulsion energy between bacteria and the glass surface was calculated by eq. S7[1,8]:

$$G_{EDL} = \pi\varepsilon_0\varepsilon_r a_b \left\{ 2\xi_b\xi_s \ln\left[\frac{1+\exp(-\kappa h)}{1-\exp(-\kappa h)}\right] + (\xi_b^2 + \xi_s^2)\ln\left[1-\exp(-2\kappa h)\right] \right\} \tag{S7}$$

where κ^{-1} is the thickness of the electrical double layer (EDL, nm) as calculated by the Guoy-Chapman theory with C and z being the molar bulk concentration and the charge number of the electrolytes[2] (eq. S8).

$$\kappa^{-1} = \left[3.29 z C^{1/2} \right]^{-1} \tag{S8}$$

For a 10 mM and a 100 mM phosphate buffer (PB) solution, a κ^{-1} of 2.15 nm (10 mM PB) and κ^{-1} of 0.65 nm (100 mM PB) were calculated.

With given values of the effective Hamaker constant A_{132}, the Lifshitz-van der Waals interaction energy can be calculated by eq. S9[9]

$$G_{LW} = -\frac{A_{132}}{6} \left[\frac{2a_b(h+a_b)}{h(h+2a_b)} - \ln(\frac{h+2a_b}{h}) \right] \tag{S9}$$

The DLVO interaction force thus can be calculated with the DLVO energy divided by separating distance between bacteria and solid surface.

$$F_{\text{DLVO}} = G_{\text{DLVO}} / h \tag{S10}$$

Calculation of hydraulic shear force F_{HF}

The shear forces F_{HF} and F_{EOF}, acting on a bacterium located at h_s depend on the velocity of the hydraulic (V_{HF}) and the electroosmotic (V_{EOF}) water flow and can be calculated by eq. S11:

$$F_{HF} = F_d^* \times 6\pi\eta a V_{HF} \tag{S11}$$

Where η is the viscosity of the liquid ($\eta = 3.19$ kg m^{-1} h^{-1}), F_d^* is a function of the radius a of a sphere (for simplicity we presume bacterial cells to be spheres) and the distance of the center of the sphere to the collector surface.

Estimation of the bacterial coverage of attached bacterial cells on sensor

The coverage of bacteria on sensor surface was estimated using eq. S12,

$$Coverage = \frac{\pi \times r_a \times r_b \times N_{Cell}}{\pi r_{sensor}^2} \times 100\% \tag{S12}$$

where N_{cell} is the cell number deposited on the sensor surface, r_a the length radius of bacteria (0.57 µm), r_b the width radius of bacteria (0.31 µm), r_{sensor} the radius of the sensor (7 mm). N_{cell} was observed from microscope counting with the method described in the main text.

Code for ImageJ automatic counting of cell number in images taken with Hemacytometer under microscope.

The cell numbers on microscopy pictures were counted with ImageJ software, going through batch macro function, with code as following:

```
run("Subtract Background...", "rolling=12 light");
run("Color Balance...");
setMinAndMax(235, 255);
run("Apply LUT");
run("Smooth");
setAutoThreshold("Default");
//run("Threshold...");
setThreshold(0, 151);
run("Convert to Mask");
run("Analyze Particles...", "size=20-Infinity circularity=0.00-1.00 show=Outlines display clear summarize");
```

The cell concentration of bacterial suspension injected in Hemacytometer (derived as described in the main text) was calculated with the cell number derived from ImageJ according to the protocol (cf. http://hausserscientific.com /products/hausser_bright_line.html).

Table S1. Contact angles of water (Θ_w), formamide (Θ_f), methylene iodide (Θ_m), and zeta potentials (ζ_{bac}) of *P. putida* KT2440 and *P. fluorescens* LP6a measured in 10 mM, 50 mM, and 100 mM phosphate buffer.

	Θ_w (degree)	Θ_f (degree)	Θ_m (degree)	ζ_{bac} 10 mM (-mV)	ζ_{bac} 50 mM (-mV)	ζ_{bac} 100 mM (-mV)
Silica	21 ± 2	40 ± 5	56 ± 4	-21 ± 2	-12 ± 1	-8 ± 1
P. putida KT2440	70 ± 3	64 ± 7	57 ± 2	-30 ± 1	-14 ± 2	-11 ± 1
P. fluorescens Lp6a	46 ± 3	55 ± 4	56 ± 2	-53 ± 2	-43 ± 2	-36 ± 3

Table S2. Calculated DLVO force (F_{DLVO}) and distances of the secondary minimum distances of *P. putida* KT2440 and *P. fluorescens* LP6a above the QCM-D silica sensor surface in 10 *m*M, 50 *m*M, and 100 *m*M phosphate buffer.

	P. putida KT2440			*P. fluorescens* LP6a		
	10 *m*M	50 *m*M	100 *m*M	10 *m*M	50 *m*M	100 *m*M
Distance of secondary minimum (*n*M)	18.8	6.0	3.2	20.6	6.7	4.1
F_{DLVO} at secondary minimum (pN)	0.15	1.5	3.3	0.15	1.4	2.3

Table S3. Effect of electric field strength (E) on attached cell density (η_c) of *P. putida* KT2440 and *P. fluorescens* LP6a cells on the QCM-D quartz sensor surface (surface area: 1.54 cm^2) as determined by microscopic counting.

	E (V cm^{-1})	*P. putida* KT2440 η_c ($\times 10^6$ Cells/cm^2)	Coverage (%)	*P. fluorescens* LP6a η_c ($\times 10^6$ Cells/cm^2)	Coverage (%)
10 *m*M	0.0	2.47	1.37	1.21	0.67
	0.5	1.15	0.64	1.35	0.75
	1.0	1.21	0.67	0.91	0.50
	2.0	0.93	0.52	0.72	0.40
50 *m*M	0.0	1.82	1.93	2.86	1.59
	0.5	2.14	1.69	0.86	0.48
	1.0	3.48	1.19	0.99	0.55
	2.0	3.55	1.01	0.83	0.46
100 *m*M	0.0	4.07	2.26	3.25	1.80
	0.5	4.76	2.64	1.94	1.07
	1.0	5.01	2.78	1.29	0.71
	2.0	9.94	5.52	1.14	0.63

Table S4. Linear correlation of Δf_5 and ΔD_5 ranges with coefficients of determination R^2 for bacterial strains *P. putida* KT2440 and *P. fluorescens* LP6a at DC electric field strengths of $E = 0$, 0.5, 1.0, or 2.0 V cm^{-1}.

Bacterial strain	*P. putida* KT2440				*P. fluorescens* LP6a			
Electric field strength (V cm^{-1})	0	0.5	1	2	0	0.5	1	2
R^2	0.98	0.98	0.99	0.99	0.98	0.96	0.90	0.07

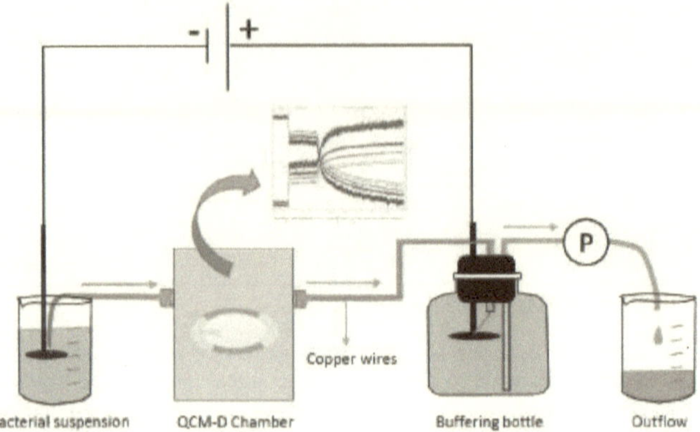

Figure S1. Schematic view of QCM-D setup to investigate electrokinetic effects on bacterial deposition. Bacterial suspension was driven through the QCM-D system under pump pressure following the direction of the blue arrows. The sealed buffering bottle was added to make sure the copper wires can be connected to the power supply without and loss of pump pressure. Cooper wires were inserted until 2 mm from the edge of the QCM-D chamber.

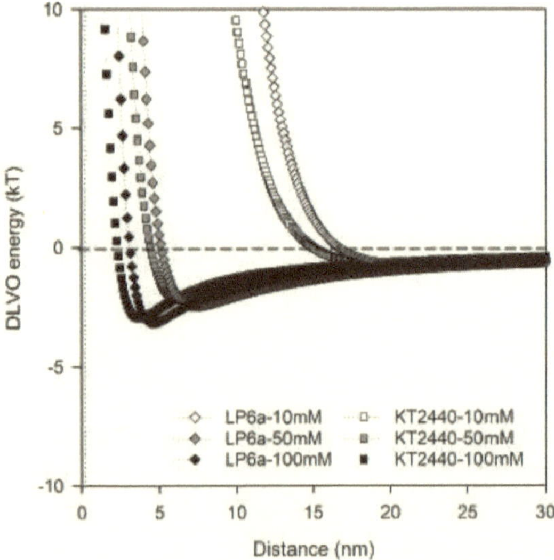

Figure S2. Calculated DLVO energy profile of *P. putida* KT2440 (square symbols) and *P. fluorescens* LP6a (diamond symbols) in 10 mM (open), 50 mM (grey), and 100 mM (black) phosphate buffer.

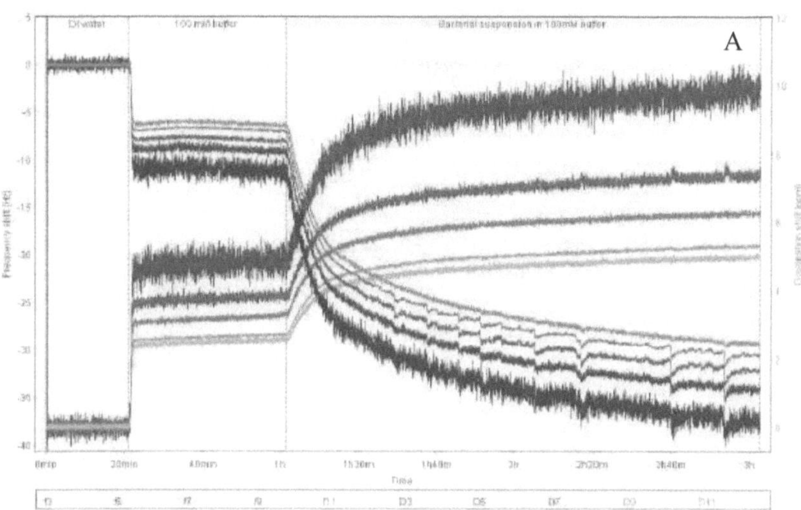

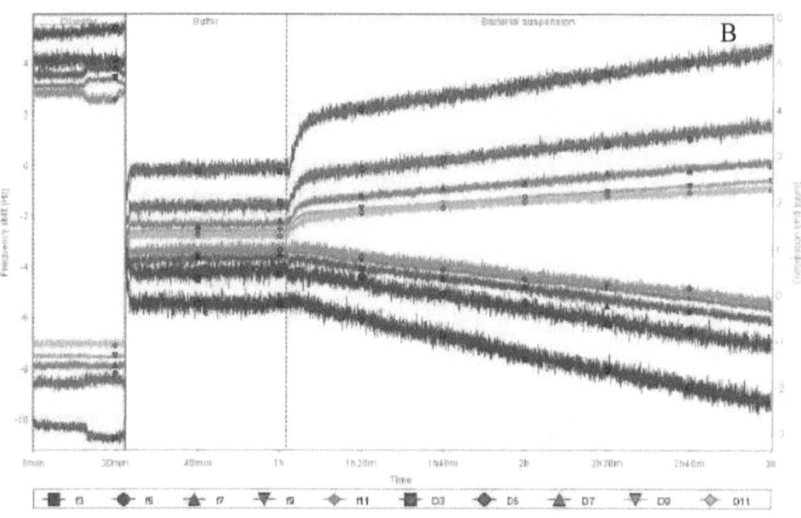

Figure S3. Typical QCM-D recorded frequency and dissipation shifts at overtones 3, 5, 7, 9, and 11 during 120 minutes of deposition of *P. putida* KT2440 (Fig. S3A) and *P. fluorescens* LP6a (Fig. S3B) to silica sensor in 100 mM buffer at $E = 1$ V cm^{-1} at overtones. The experiments were run in three phases by pumping: 1. deionized water, 2. phosphate buffer, and 3. bacterial suspension over the sensor surface.

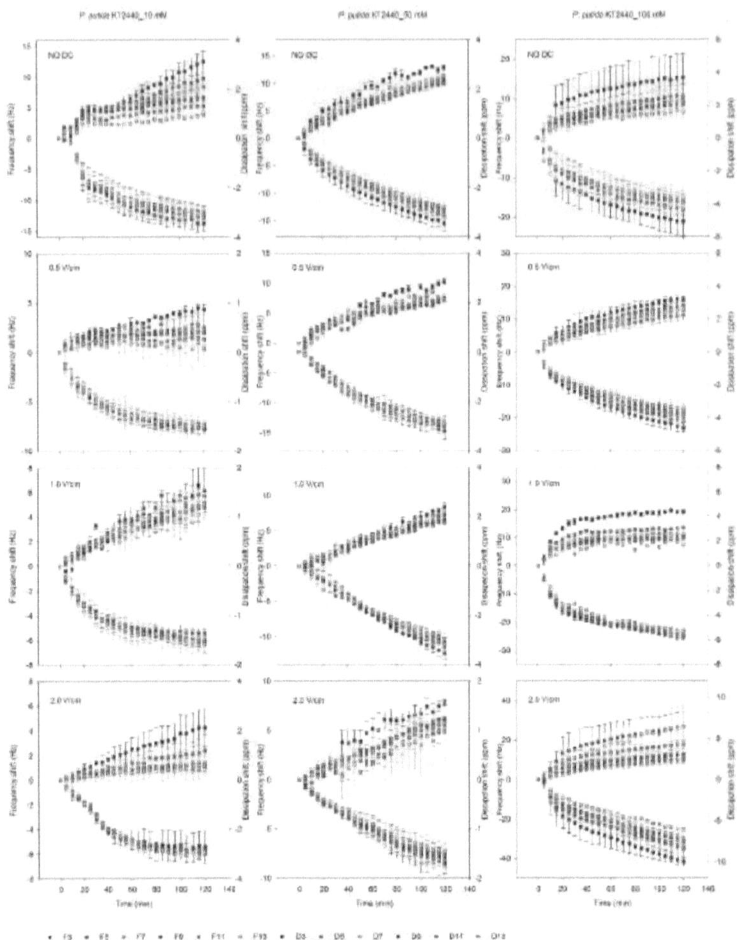

Figure S4. Δf and ΔD signals of *P. putida* KT2440 cell deposition on silica sensor at an interval of 5 min at overtones 3 to 13 (error bars are the standard deviation of triplicate experiments).

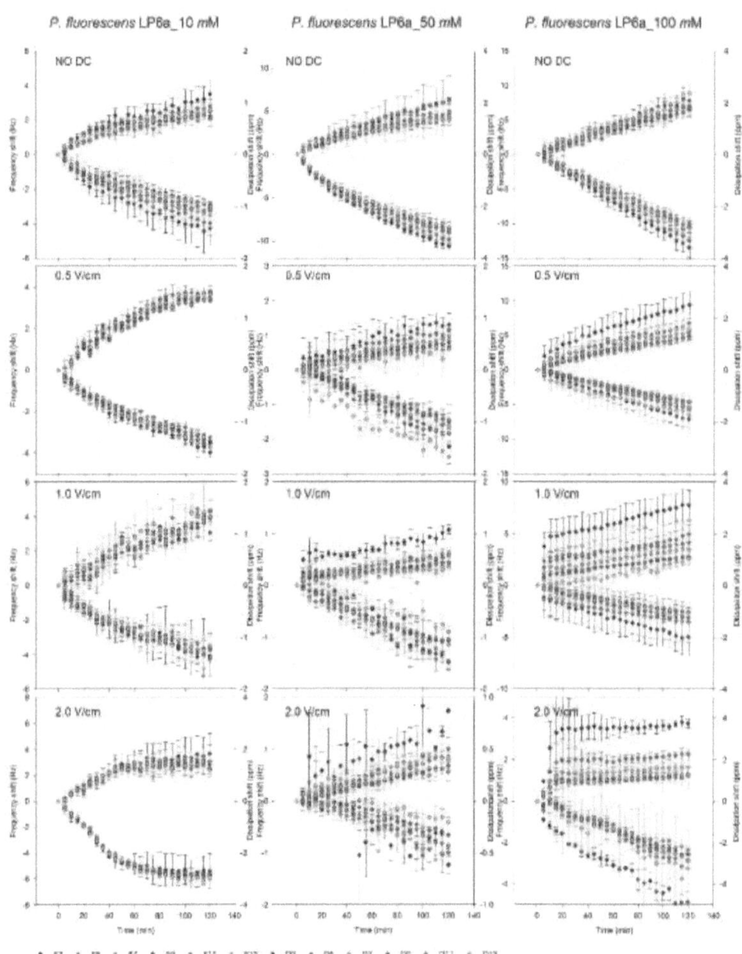

Figure S5. Δf and ΔD signals of *P. fluorescens* LP6a cell deposition on silica sensor at an interval of 5 min at overtones 3 to 13 (error bars are the standard deviation of triplicate experiments).

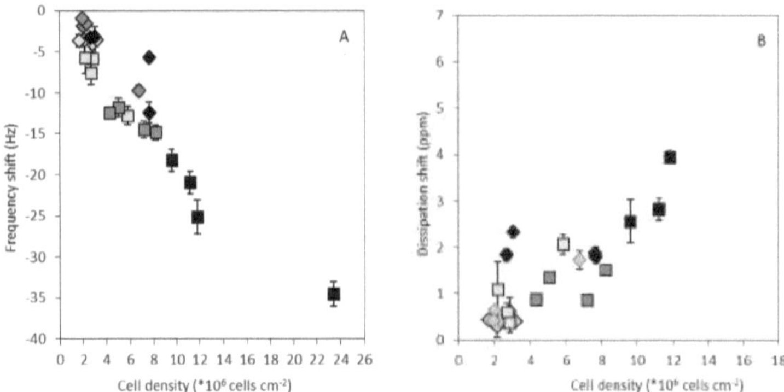

Figure S6. Correlation of Δf_5 (panel A) and ΔD_5 (panel B) with attached cell density (η_c) on the silica sensor surface after 120 minutes observed by microscopy counting (error bars represent for standard deviations of triplicate experiments). The data points include two bacterial strains KT2440 (square symbols) and LP6a (diamond symbols) in 10 mM (gray), 50 mM (dark gray), and 100 mM (black) PB.

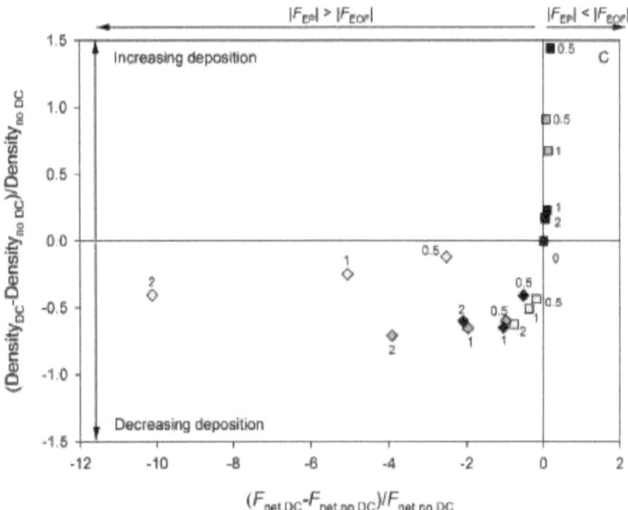

Figure S7. Relative change of the cell density and relative change of DC-induced net forces acting on a bacterium placed at the secondary minimum. All plots reflect data after two hours of deposition of *P. putida* KT2440 (squares) and *P. fluorescens* LP6a (diamonds) exposed to electrolyte concentration of either 10 mM (gray symbols), 50 mM (dark gray symbols), and 100 mM (black symbols) and DC electric field strengths of either E = 0, 0.5, 1.0, or 2.0 V cm^{-1}.

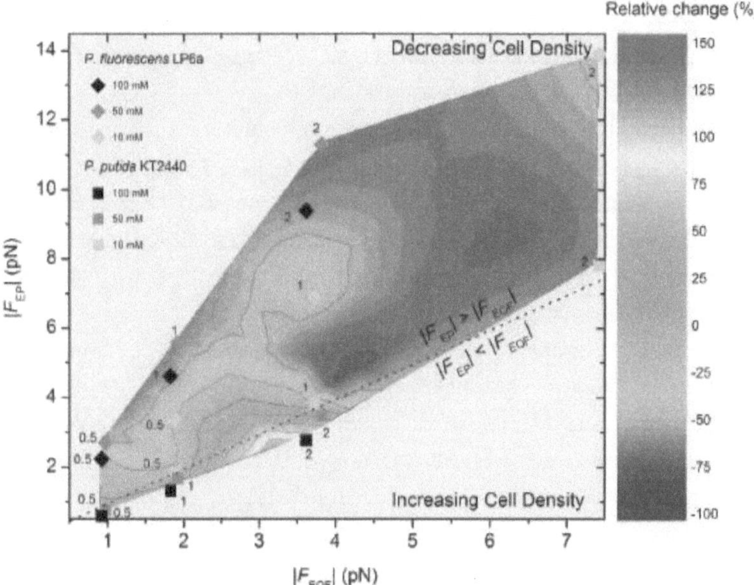

Figure S8. Calculated effects of the electroosmotic shear force $|F_{EOF}|$ and the electrophoretic drag force $|F_{EP}|$ acting on a bacterium placed at the secondary minimum on the bacterial cell density on the sensor surface relative to DC-free controls in percentage. The plots reflect data after two hours of deposition of *P. putida* KT2440 (squares) and *P. fluorescens* LP6a (diamonds) exposed to electrolyte concentration of either 10 *m*M (light gray symbols), 50 *m*M (dark gray symbols), and 100 *m*M (black symbols) and DC electric field strengths of either $E = 0$, 0.5, 1.0, or 2.0 V cm^{-1}.

References

(1)   Boks, N. P.; Norde, W.; van der Mei, H. C.; Busscher, H. J. Forces Involved in Bacterial Adhesion to Hydrophilic and Hydrophobic Surfaces. *Microbiology* **2008**, *154* (10), 3122–3133. https://doi.org/10.1099/mic.0.2008/018622-0.

(2)   Sharma, P. K.; Hanumantha Rao, K. Adhesion of Paenibacillus Polymyxa on Chalcopyrite and Pyrite: Surface Thermodynamics and Extended DLVO Theory. *Colloids Surf. B Biointerfaces* **2003**, *29* (1), 21–38. https://doi.org/10.1016/S0927-7765(02)00180-7.

(3)   Van Oss, C. J.; Chaudhury, M. K.; Good, R. J. Interfacial Lifshitz-van Der Waals and Polar Interactions in Macroscopic Systems. *Chem. Rev.* **1988**, *88* (6), 927–941. https://doi.org/10.1021/cr00088a006.

(4)   Martin, R. E.; Bouwer, E. J.; Hanna, L. M. Application of Clean-Bed Filtration Theory to Bacterial Deposition in Porous Media. *Environ. Sci. Technol.* **1992**, *26* (5), 1053–1058.

(5)   Vilinska, A.; Rao, K. H. Surface Thermodynamics and Extended DLVO Theory of Leptospirillum Ferrooxidans Cells' Adhesion on Sulfide Minerals. *Min. Metall. Explor.* **2011**, *28* (3), 151–158. https://doi.org/10.1007/BF03402248.

(6)   Fowkes, F. M. Attractice Forces at Interfaces. *Ind. Eng. Chem.* **1964**, *56* (12), 40–52. https://doi.org/10.1021/ie50660a008.

(7)   Brown, D. G.; Jaffé, P. R. Effects of Nonionic Surfactants on the Cell Surface Hydrophobicity and Apparent Hamaker Constant of a *Sphingomonas* Sp. *Environ. Sci. Technol.* **2006**, *40* (1), 195–201. https://doi.org/10.1021/es051183y.

(8)   Van Oss, C. J.; Docoslis, A.; Wu, W.; Giese, R. F. Influence of Macroscopic and Microscopic Interactions on Kinetic Rate Constants. *Colloids Surf. B Biointerfaces* **1999**, *14* (1–4), 99–104. https://doi.org/10.1016/S0927-7765(99)00028-4.

(9)   Van Oss, C. J.; Good, R.; Chaudhury, M. The Role of van Der Waals Forces and Hydrogen Bonds in "Hydrophobic Interactions" between Biopolymers and Low Energy Surfaces. *J. Colloid Interface Sci.* **1986**, *111* (2), 378–390. https://doi.org/10.1016/0021-9797(86)90041-X.

5. Electrokinetic Effects on Matrix-Contaminant Interactions

5.1 Electrokinetic Effects on the Interaction of Phenanthrene with Geo-Sorbents

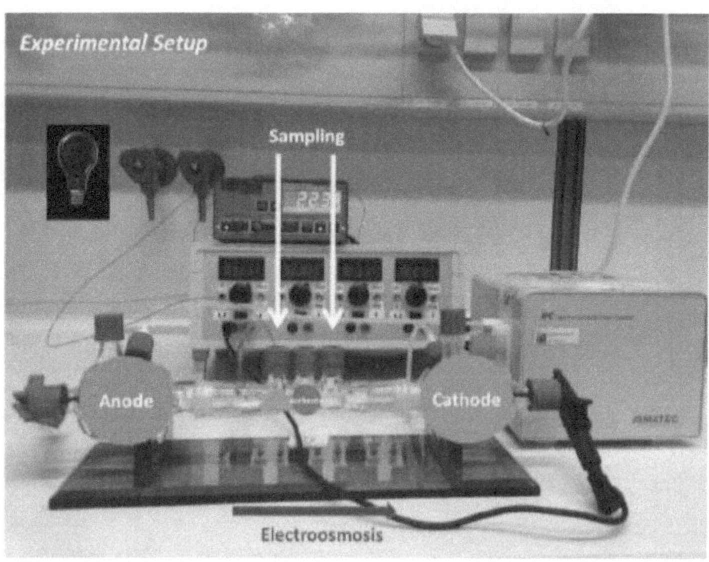

Chemosphere 241 (2020) 125164

Contents lists available at ScienceDirect

Chemosphere

journal homepage: www.elsevier.com/locate/chemosphere

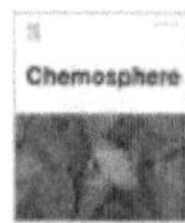

Electrokinetic effects on the interaction of phenanthrene with geo-sorbents

Yongping Shan [1], Jinyi Qin [1,2], Hauke Harms, Lukas Y. Wick [*]

Helmholtz Centre for Environmental Research - UFZ, Department of Environmental Microbiology, 04318, Leipzig, Germany

HIGHLIGHTS

- DC electric fields strongly change PAH interactions with (non-electrode) sorbents.
- DC effects correlate to the electroosmotic flow velocity and $\Delta G°$ of PAH sorption.
- DC fields increase PAH sorption to and reduce PAH desorption from strong sorbents.
- DC fields limit PAH sorption to and promote PAH desorption from weak sorbents.

GRAPHICAL ABSTRACT

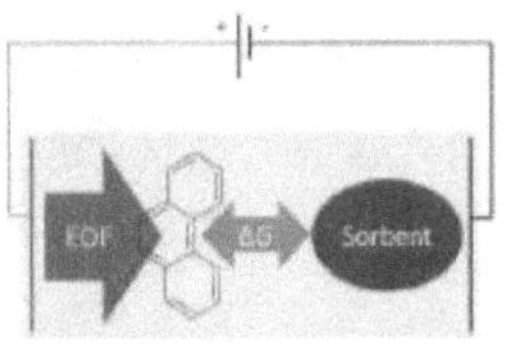

ARTICLE INFO

Article history:
Received 20 August 2019
Received in revised form
16 October 2019
Accepted 20 October 2019
Available online 25 October 2019

Handling Editor: X. Cao

Keywords:
Phenanthrene
Electrokinetics
Electroosmotic flow
Sorption thermodynamics

ABSTRACT

Interactions with solid matrices control the persistence and (bio-)degradability of hydrophobic organic chemicals (HOC). Approaches influencing the rate or extent of HOC interactions with matrices are thus longed for. When a direct current (DC) electric field is applied to a matrix immersed in an ionic solution, it invokes transport processes including electromigration, electrophoresis, and electroosmotic flow (EOF). EOF is the surface charge-induced movement of pore fluids. It has the potential to mobilize uncharged organic contaminants and, hence, to influence their interactions with sorbing geo-matrices (i.e. geo-sorbents). Here, we assessed the effects of weak DC electric fields on sorption and desorption of phenanthrene (PHE) in various mineral and carbonaceous geo-sorbents. We found that DC fields significantly changed the rates and extent of PHE sorption and desorption as compared to DC-free controls. A distinct correlation between the Gibbs free energy change ($\Delta G°$) and electrokinetic effects such as the EOF velocity was observed: in case of mineral sorbents EOF limited (or even inhibited) PHE sorption and increased its desorption. In strongly sorbing carbonaceous geo-sorbents, however, EOF significantly increased the rates of PHE sorption and reduced PHE desorption by > 99% for both activated charcoal and exfoliated graphite. Based on our findings, an approach linking $\Delta G°$ and EOF velocity was developed to estimate DC-induced PHE sorption and desorption benefits on mineral and carbonaceous sorbents. We conclude that such kinetic regulation gives rise to future technical applications that may allow modulating sorption processes e.g. in response to fluctuating sorbate concentrations in contaminated water streams.

© 2019 Elsevier Ltd. All rights reserved.

* Corresponding author. Helmholtz Centre for Environmental Research – UFZ, Department of Environmental Microbiology, Permoserstrasse 15, 04318, Leipzig, Germany.
  E-mail address: lukas.wick@ufz.de (L.Y. Wick).
[1] Both authors contributed equally to the manuscript.
[2] Present address: School of Civil Engineering, Chang'an University, Xi'an 710054, P.R. China.

https://doi.org/10.1016/j.chemosphere.2019.125164

Y. Shan et al. / Chemosphere 242 (2020) 125161

1. Introduction

Interactions with solid geo-matrices are key drivers of the persistence of hydrophobic organic chemicals (HOC) and, hence, control their fate as well as the exposure of environmental and human receptors (Harms et al., 2017a). Various studies have shown that the sequestration of hydrophobic chemicals in the solid phase significantly reduces HOC bioavailability and biodegradation (Harms et al., 2017a). Three potentially rate-limiting steps may influence the sorption of a chemical to and its release from geo-matrices, respectively: (i) diffusion of the chemical within the molecular nano-porous network, (ii) pore or surface diffusion in aggregated geo-matrices, and (iii) diffusion of the sorbate across an aqueous boundary layer surrounding sorbent particles. As a consequence of progressive binding, residual hydrophobic chemical may become less leachable and thus less efficiently available for microbial degradation (Johnsen et al., 2005). The sorption and release of HOC in various sorbents can be investigated by kinetic (Ahn et al., 2009; Ho et al., 2000; Morelis and van Noort, 2008) and thermodynamic (e.g., Gibbs free energy of sorption, ΔG°) approaches (Doke and Khan, 2013; Huang and Weber, 1997). Electrokinetic approaches (Qin et al., 2015a, 2015b; Shan et al., 2018) (typically using DC fields of an electric field strength of $X = 0.2$–2.0 V cm^{-1}) have shown high potential to mobilize otherwise poorly mobilizable hydrophobic organic pollutants in matrices of low permeability (Jeon et al., 2010; Kim et al., 2006; Li et al., 2000; Pham et al., 2009; Reddy, 2010; Saichek and Reddy, 2003; Shapiro and Probstein, 1993). When an electric field is applied to a matrix immersed in an ionic solution, it invokes transport processes including electromigration, electrophoresis, and electroosmotic flow (EOF). Electromigration and electrophoresis refer to the transport of charged molecules and particles to electrodes of opposite charge. EOF is the surface charge-induced movement of pore fluids in an electric field usually moving from the anode toward the cathode (Rice and Whitehead, 1965). It originates from the enrichment of ions in the so-called electric double layer near a surface and is particularly effective in fine-grained materials where meso- and micro-pores dominate. These are situations where low hydraulic flow is extremely small and molecular diffusion may limit the access of sorbates to and the release of sorbates from smaller pores (Reddy and Saichek, 2004; Shi et al., 2008). Electroosmotic perfusion induces efficient liquid flow in inter- and intra-particle network pore channels (Tallarek et al., 2001) and, hence, increase release rates and natural attenuation of HOC at locations where pump and treat approaches may be inadequate (Kim et al., 2007; Reddy et al., 2009; Reddy and Saichek, 2004) or energetically ineffective (Hassan et al., 2015). EOF can thus be applied for the dispersal and separation of uncharged entities or the dewatering of matrices (Hoshyargar et al., 2018; Li et al., 2018). Contrary to the parabolic velocity profile of pressure-driven hydraulic flow in a pore, the velocity profile of EOF is quasi planar beginning at the so-called electrical double layer located a few nanometers above the surface. It thus likely arises at scales relevant for chemical-sorbent interactions. This effect is, for instance, used in capillary electrochromatography (CEC) where EOF (rather than pressure-driven-flow such as in HPLC) is used to effectively separate uncharged solutes between a mobile and a stationary phase (Vallano and Remcho, 2000). Recent work showed that DC fields increased PHE sorption rates in carbonaceous exfoliated graphite sevenfold and reduced the PHE desorption rate by > 99% (Qin et al., 2015a). This was discussed as a result of electroosmotic perfusion of PHE to pores that contribute most of the sorption sites, but are difficult to access in the absence of EOF by molecular diffusion only. Still scarce mechanistic information exists on the impact of DC on the sorption and desorption of hydrophobic chemicals with geo-matrices. In our study we tested whether electrokinetic phenomena are able to change interactions of the common oil contaminant (Bansal and Kim, 2015) phenanthrene (PHE) with mineral (zeolites, aluminum oxides, silicates) and carbonaceous (activated charcoal, exfoliated graphite) sorbents as compared to DC-free controls. In particular, we assessed to which degree (i) DC electric fields influence the sorption and desorption of PHE in model sorbents of differing sorption strength, (ii) DC-induced benefits of PHE sorption/desorption correlate with Gibbs free enthalpy (ΔG°), and (iii) EOF may explain DC-induced benefits of PHE-geo-matrix interactions.

2. Material and methods

2.1. Reagents and sorbents

One hundred mg PHE (purum > 97.0% GC; Fluka, Germany) were stirred in 100 mL of methanol in a light-protected volumetric flask for 1 h. One mL of the methanolic PHE solution was then added to 1 L of either a sterile 1, 10 or 100 mmol L^{-1} potassium phosphate buffer (PB; i.e. K$_2$HPO$_4$, KH$_2$PO$_4$) at pH ~ 7, shaken for 24 h under light exclusion and stored at 4 °C until use. Except for the exfoliated graphite (EG), which was prepared as described earlier (Moustafa, 2009; Qin et al., 2015a), all sorbents [silica gel 40 Å, 60 Å, 100 Å (Davisil), aluminum oxide (Al$_2$O$_3$; Si/Al = 0), zeolite NaY (Si/Al = 5), zeolite 13X (Si/Al = 1.4) and activated charcoal (AC)] were purchased from Sigma-Aldrich Chemie GmbH, Germany. The mineral sorbents were cleaned by exposure to 500 °C for 4 h in a muffle furnace and subsequent storage in a desiccator. For desorption experiments the sorbents were spiked with PHE as described earlier (Preglis et al., 2007); shortly, 10 mg of the sorbents were added to 10 mL of hexane containing 1 mg mL^{-1} of PHE in a tightly sealed vial and sonicated for 1 h in an ultrasonic bath (Sonorex Super RK255/H, Bandelin Electronic GmbH). Thereafter, the vial was opened and the hexane allowed evaporating at 60 °C for 10 min. After loading, all of the sorbents were stored at 5 °C in a closed vial in the dark.

2.2. Analytical methods

Aqueous samples with dissolved PHE were analyzed by high performance liquid chromatography (HPLC) (Shimadzu Class-VP) on an RP-18 column (Nucleosil 100-5 C18 4 mm ID) using an isocratic mobile phase (MeOH/water (90:10 v/v); flow: 1 mL min^{-1}) and UV detection at 250 nm. The physicochemical surface characteristics (BET surface, pore size and the zeta potential (ζ)) of the sorbents were characterized by Doppler electrophoretic light scattering analysis, and BET analysis (cf. Fig. S1).

2.3. Electrokinetic apparatus and running conditions

The electrokinetic apparatus used for the sorption and desorption was composed of two electrode compartments and three central chambers, (cf. Fig. S2) as described earlier by Qin et al. (2015a). The apparatus was mounted horizontally, filled with either 1, 10, 50, or 100 mmol L^{-1} electrolyte, and connected to a PowerPac (P333, Szczecin, Poland) to produce an electric field of an electric field strength X of 1.8 V cm^{-1} and resulting currents I of 3 ± 0.4, 14 ± 2, 30 ± 3 mA for 1, 10, and 100 mmol L^{-1} of electrolyte, respectively. The electrolyte was circulated from the anode to the cathode by a peristaltic pump (ISM 935, Ismatec, Glattbrugg, Switzerland) with sterilized Teflon at 26.4 mL h^{-1} (bed flow velocity; 0.004 cm s^{-1}). Experiments in the absence of electric field were conducted as controls.

2.4. Kinetics of PHE geo-sorbent interactions

2.4.1. Desorption

Desorption of PHE from PHE-loaded sorbents was quantified in presence and absence of a DC electric field ($X = 1.8\ V\ cm^{-1}$) in 10, 50 and 100 mmol L^{-1} electrolyte solutions. The central of the three chambers (cf. Fig. S2) was packed with dry, PHE-spiked sorbent prepared as described by (Puglisi et al., 2007); either 1 g of a mineral sorbent (silica 40 Å, 60 Å, 100 Å, Al_2O_3, zeolite NaY, zeolite 13X), 0.5 g of AC, or 70 mg of EG. The apparatus then was filled with 200 mL PB solution (denoted as $t = 0$ h) and continuously flushed overnight (11 h) for system equilibration prior to the application of the DC electric field. Thereafter samples of 1 mL were taken from chamber (3) (cf. Fig. S2), transferred each to 1.5 mL glass vials and the PHE content analyzed by HPLC. No significant changes of the temperature and the pH of the electrolyte were observed in either of the ionic strengths of the electrolyte used. All experiments were performed in three independent replicates.

2.4.2. Sorption

Sorption of PHE to sorbents was quantified in presence and absence of a DC electric field ($X = 1.8\ V\ cm^{-1}$) in 1, 10 and 100 mmol L^{-1} electrolyte solutions. The chamber (4) (cf. Fig. S2) was packed with dry, cleaned sorbent as described above, the apparatus was filled with 200 mL electrolyte solution and then continuously flushed with PHE containing electrolyte ($C_e = 400\ \mu g\ L^{-1}$) at a rate of 26.4 mL h^{-1} under DC and DC-free conditions. After a stabilization time of 5 min (denoted as $t = 0$ h), samples of 1 mL were taken at given intervals from both chambers (1) and (3) next to the central chamber (4): (cf. Fig. S2), transferred to 1.5 mL glass vials and the PHE contents were analyzed by HPLC. All experiments were performed in three independent replicates.

2.5. Determination of sorption isotherms

Triplicate batch experiments of PHE sorption on all sorbents were performed separately either at $5 \pm 2\ °C$, $25 \pm 2\ °C$, and $35 \pm 2\ °C$ in 10 and 100 mmol L^{-1} PB. Isothermal sorption experiments were performed according to a standard protocol (Huang and Weber, 1997; Su et al., 2006; Zhao et al., 2014) in 30 mL (mineral sorbents) or 200 mL (carbonaceous sorbents) glass vials that were sealed with a Teflon-coated butyl rubber septum crimp cap. Activated charcoal was ground into fine particles (diameter $14 \pm 3\ \mu m$, $n = 20$) to shorten the time needed to reach sorption equilibrium (James et al., 2005). The solid-to-solution ratios (w/v) were 1:20 (g mL^{-1}) for PHE sorption to mineral sorbents and 1:8000 (g mL^{-1}) for carbonaceous sorbents, respectively. Dissolved PHE concentrations of 100, 200, 300, 400, and 500 $\mu g\ L^{-1}$ were used for mineral sorbent while 10, 20, 30, 40, 50, 60, 70, and 80 mg L^{-1} were used for carbonaceous sorbents (0.1%, 0.2%, 0.3%, 0.4%, 0.5%, 0.6%, 0.7%, and 0.8% MeOH was added as co-solvent accordingly). The glass vials were horizontally shaken in a reciprocal shaker (3016, GFL, Germany) at 12 rpm for 7 days. All experiments were performed in three independent replicates. Aqueous PHE concentrations were quantified by HPLC as mentioned above. Sorption isotherms were approximated using Freundlich adsorption isotherm (eq. (1)) (Ai and Jiang, 2012; Duan and Naidu, 2013; Zhang et al., 2010).

$$\log q_e = \log K_F + n \log C_e \tag{1}$$

where q_e is the equilibrium concentration of PHE adsorbed to sorbents, C_e is the dissolved PHE equilibrium concentration, n is the Freundlich exponent (a measure of sorption linearity) and K_F is the Freundlich isotherm constant ($\mu g\ kg^{-1}\ (L\ \mu g^{-1})^n$). The distribution coefficient K_d at equilibrium was determined by $K_d = q_e/C_e$ ($L\ g^{-1}$). The specific surface-normalized distribution coefficient K_d^* can be further calculated by dividing K_d by the specific surface area ($m^2\ g^{-1}$) of the sorbents.

2.6. Calculation of sorbed and desorbed PHE fractions

$\Gamma_{des,t}$ (%) and $\Gamma_{sor,t}$ (%) refer to normalized time-dependent fractions of PHE in the sorbent in desorption and sorption experiments, respectively. They were calculated from PHE inflow (C_i) and outflow concentrations (C_e) of the reactor chamber, the electrolyte volume flushed (V: L), and the initial PHE load (M_0: mg) in the sorbent and the maximum amount of PHE that can be loaded on clean sorbent in the column (M_t: mg), respectively (eqs. (2) and (3))

$$\Gamma_{des\ t} = \frac{M_0 - \int_0^t C_e\ dV}{M_0} \tag{2}$$

$$\Gamma_{sor\ t} = \frac{\int_0^t C_i dV - \int_0^t C_e dV}{M_b} \tag{3}$$

The relative influence of DC electric fields on PHE desorption ($\Delta\Gamma_{des,t}$) and sorption ($\Delta\Gamma_{sor,t}$) at a given time can be calculated from eqs. (4) and (5), where subscripts denote the absence and presence of the electric field.

$$\Delta\Gamma_{des\ t} = \Gamma_{des\ noDC\ t} - \Gamma_{des\ DC\ t} \tag{4}$$

$$\Delta\Gamma_{sor\ t} = \Gamma_{sor\ noDC\ t} - \Gamma_{sor\ DC\ t} \tag{5}$$

2.7. Thermodynamics of PHE geo-sorbent interactions

The standard Gibbs free energy of sorption ($\Delta G°$) relates to standard sorption enthalpy ($\Delta H°$) and sorption entropy changes ($\Delta S°$) by eq. (6).

$$\Delta G° = \Delta H° - T\Delta S° \tag{6}$$

$\Delta G°$ can be estimated according to the following equation at $T = 298\ K$:

$$\Delta G° = -RT\ \ln K_c \tag{7}$$

K_c is the equilibrium constant; it is dimensionless and based on the Freundlich isotherm K_F (at 298 K) and the water density (ρ) 1000 g L^{-1}, and can be calculated using eq. (8) (Ghosal and Gupta, 2015; Kopinke et al., 2018; Tran et al., 2017):

$$K_c = \frac{K_F \rho}{1000}\left(\frac{10^6}{\rho}\right)^{(1-n)} \tag{8}$$

$\Delta H°$ can be estimated using the van't Hoff equation by substituting eq. (7) to eq. (6) (Kopinke et al., 2018; Tran et al., 2017):

$$\ln K_c = \frac{-\Delta H°}{R} \times \frac{1}{T} + \frac{\Delta S°}{R} \tag{9}$$

The $\Delta H°$ (kJ mol^{-1}) is a measure of the enthalpy change (isosteric heat) involved in the transfer of solute from the reference state to the sorbed state at a given solid-phase concentration. R is the universal gas constant (8.314×10^{-3} kJ $mol^{-1}\ K^{-1}$) and T is temperature in Kelvin. The values of $\Delta H°$ can be estimated by the slope and intercept of a plot of $\ln K_c$ versus $1/T$, and $\Delta S°$ can be calculated by ΔH° and ΔG° according to eq. (6).

2.8. Approximation of the EOF velocity

The electroosmotic (EOF) flow velocity in an intra-particle pore of radius r ($V_{EOF,r}$) can be calculated by $V_{EOF,max}$ and a function of κr (i.e., $f(\kappa r)$) (Rice and Whitehead, 1965) as detailed by eqs. (10)–(12).

$$V_{EOF,r} = V_{EOF,max} \cdot f(\kappa r) \tag{10}$$

$$V_{EOF,max} = - \frac{\varepsilon_r \cdot \varepsilon_D \cdot X \cdot \zeta}{\eta} \tag{11}$$

$$f(\kappa r) = \left(1 - \frac{2I_1(\kappa r)}{\kappa r I_0(\kappa r)}\right) \tag{12}$$

Here I_0, I_1 are the zero and first-order modified Bessel functions, κ^{-1} the thickness of the electrical double layer (EDL, nm) calculated by the Guoy-Chapman theory with C and z the molar bulk concentration and the charge number of the electrolytes, respectively (Sharma and Rao, 2003) (eq. (13)).

$$\kappa^{-1} = \left[3.29zC^{1/2}\right]^{-1} \tag{13}$$

The term κr reflects the ratio of the pore radius (r) to the thickness of the double layer. Equation (10) is a simplified expression of the Navier-Stokes equation (Cummings et al. 2000), where, ε_r is the dielectric constant of water (78.5), ε_D (8.85×10^{-12} F m^{-1}) is the vacuum permittivity, ζ is the actual zeta potential of the solid surface at the experimental conditions, and X is the electric field strength applied.

3. Results

3.1. Characterization of sorbent properties

Physico-chemical properties including specific surface area, average pore size, and zeta potential were analyzed for all eight sorbents (Table 1). All sorbents had average pore sizes of 2–10 nm and were of a high specific surface area (>190 m^2 g^{-1}), with AC, EG, silica 40 Å and zeolite 13X expressing specific surface area >500 m^2 g^{-1}. Although the carbonaceous materials had lower charges, all sorbents exhibited clearly negative zeta potentials at all ionic strengths tested (Table 1). Silica 40 Å, silica 60 Å, and silica 100 Å exhibited similar zeta potentials in all the electrolyte concentrations. The equilibrium sorption capacities were reflected by the surface normalized partition coefficients K_d' ranging from 1×10^{-6} mL cm^{-2} (mineral sorbents) to 0.148×10^{-3} mL cm^{-2} for carbonaceous sorbents. Minor effects of the ionic strength on K_d' for 10 mmol L^{-1} and 100 mmol L^{-1} at 25 °C were found (Table 1).

3.2. Effect of the DC field on PHE-sorbent interactions

In order to assess the effect of weak DC electric fields on PHE-sorbent interactions, all sorbents were exposed to a DC electric field of $X = 1.8$ V cm^{-1} at fixed bed conditions and the extent and the rates of PHE sorption and desorption compared to DC-free controls. Clear DC-field effects on both PHE sorption and desorption were observed as evidenced by time dependent relative PHE fractions remaining in the sorbents in desorption $\Gamma_{des,t}$ (%) and sorption $\Gamma_{sor,t}$ (%) experiments (Figs. 1 and S3). In mineral sorbents the DC field stimulated the PHE desorption (Fig. 1A and C) and reduced the extent of PHE sorption by 30—40% (Fig. 1B and

Table 1

Overview of the properties (BET surface area, pore size and zeta potential (ζ)) of the sorbents used in the study and their temperature-dependent Freundlich sorption isotherm parameters (K_f and n) of PHE sorption at $C_c = 400\,\mu g\,L^{-1}$ in 10 and 100 mmol L^{-1} electrolyte.

| Sorbents | Properties | | | | | | 10 mmol L^{-1} | | | | | | | 100 mmol L^{-1} | | | | | | |
| | BET Surface (m^2 g^{-1}) | Pore Size (nm) | Zeta potential ζ (mV) | | | | 5 °C | | 25 °C | | 35 °C | | 25 °C | 5 °C | | 25 °C | | 35 °C | | 25 °C |
			1 mM	10 mM	50 mM	100 mM	log K_f [a]	n	log K_f [a]	n	log K_f [a]	n	K_d' [b]	log K_f [a]	n	log K_f [a]	n	log K_f [a]	n	K_d' [b]
Silica 40 Å	675	4	-39.8	-33.3	-22	-18.5	-0.80	0.83	-1.05	0.84	-1.17	0.84	0.0049	-1.05	0.84	-1.34	0.84	-1.58	0.85	0.0026
Silica 60 Å	404	6	-35.1	-28.3	-20.3	-17	-1.48	0.83	-1.79	0.84	-1.83	0.84	0.0016	-1.45	0.84	-1.80	0.84	-1.90	0.84	0.0015
Silica 100 Å	380	10	-33.5	-25	-22.7	-17.2	-1.66	0.83	-1.83	0.84	-1.88	0.85	0.0019	-1.52	0.84	-1.83	0.84	-1.95	0.84	0.0018
Zeolite NaY	199	2	-43.5	-41.2	-30.8	-25.8	-0.55	0.43	-0.64	0.42	-0.74	0.44	0.00037	-0.45	0.42	-0.65	0.44	-0.81	0.44	0.00049
Zeolite 13X	683	2.1	-47.8	-45.2	-33.8	-28.3	-2.47	0.90	-2.77	0.91	-2.95	0.91	0.00015	-2.03	0.91	-2.30	0.91	-2.53	0.92	0.00043
Al$_2$O$_3$	782	3.5	-39.9	-35.5	-28.5	-27.6	-2.85	0.94	-3.25	1.04	-3.00	0.93	0.00092	-2.76	1	-3.02	1.00	-3.09	0.90	0.00097
AC [c]	717	2.8	n.a. [d]	-17	n.a. [d]	-11	1.54	1.96	1.4	2.65	1.20	2.71	0.136	2.16	2.01	1.84	2.16	1.61	3.44	0.148
EG [e]	525	n.a. [d]	n.a. [d]	-9	n.a. [d]	-5.6	3.67	1.96	3.14	2.12	2.76	2.14	0.133	1.62	1.93	0.74	1.98	0.69	2.09	0.138

[a] log K_f as (μg kg^{-1})(L μg^{-1})^n.
[b] surface normalized K_d' ($\zeta \times 10^{-3}$ mL cm^{-2}).
[c] activated charcoal.
[d] n.a. – not applicable.
[e] exfoliated graphite.

Y. Shan et al. / Chemosphere 242 (2020) 125163

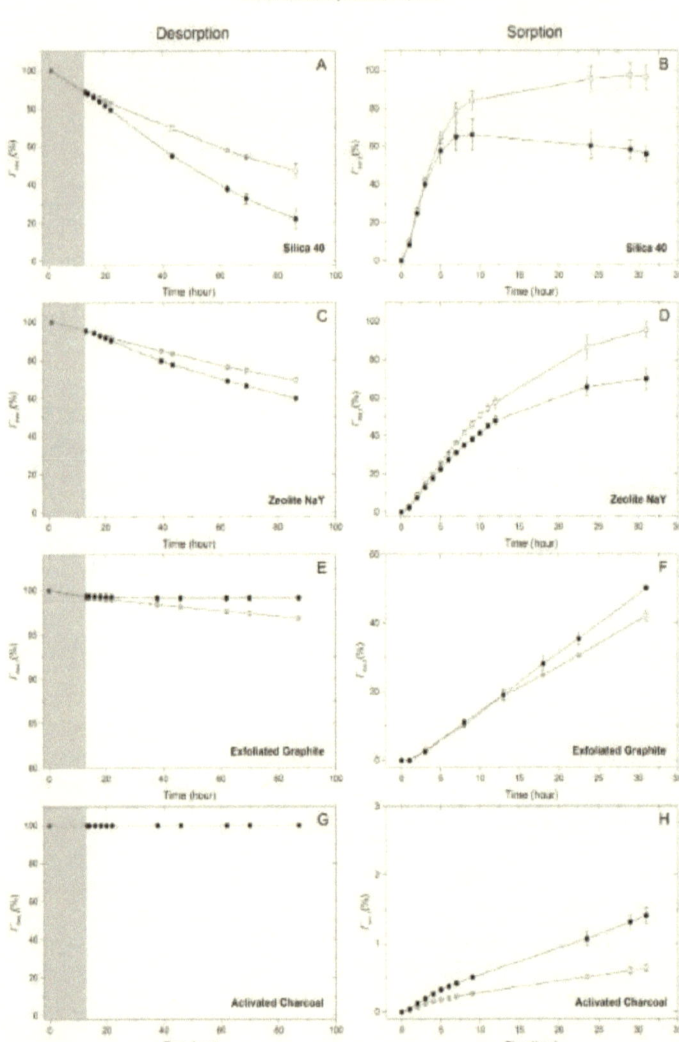

Fig. 1. Time dependent normalized PHE fractions in desorption (F_{desc}) and sorption (F_{sorp}) experiments in presence (filled symbols) and absence (open symbols) of a DC electric field using a 10 mmol L^{-1} PB electrolyte: silica 40 Å (Fig. 1A and B), zeolite NaY (Fig. 1C and D), exfoliated graphite (Fig. 1E and F), activated charcoal (Fig. 1G and H). Areas with a gray background refer to DC-free periods of the experiments. Data represent averages and standard deviations of triplicate experiments.

Y. Shan et al. / Chemosphere 242 (2020) 125161

D). By contrast, in carbonaceous EG and AC, DC field treatment increased PHE sorption (Fig. 1F and H), and limited PHE desorption from EG (Fig. 1E). No PHE release from AC was observed at any condition (Fig. 1G). As EOF velocity depends on the ionic strength of the electrolyte, PHE desorption and sorption experiments further evaluated the effect of different electrolyte ionic strengths (1, 10, 100 mmol L^{-1} PB in sorption experiments and 10, 50, 100 mmol L^{-1} PB in desorption experiments). As the electrolyte ionic strengths also may influence the sorption properties in the absence of DC, we calculated the effects of DC electric fields on PHE desorption ($\Delta \Gamma_{des,t}$) and sorption ($\Delta \Gamma_{sor,t}$) relative to DC free controls. Positive $\Delta \Gamma_{des,t}$ refers to increased desorption in presence of DC, while a negative $\Delta \Gamma_{des,t}$ refers to reduced desorption. Similarly, positive $\Delta \Gamma_{sor,t}$ denotes decreased sorption to the geo-sorbents relative to the control, while negative values refer to increased sorption. Fig. 2 depicts $\Delta \Gamma_{des,t}$ and $\Delta \Gamma_{sor,t}$ for two mineral [silica 40 Å, zeolite NaY] and two carbonaceous (AC, EG) materials exposed to various electrolyte ionic strengths: at low (1 mM) electrolyte concentrations the DC electric field showed minor or no significant change of $\Gamma_{sor,t}$ and, hence minor DC effects on PHE sorption (Figs. 2 and S4) while an increase of the electrolyte ionic strength resulted in $\Delta \Gamma_{sor,t} > 0$ (i.e., reduced sorption, Fig. 2B and D) for mineral and $\Delta \Gamma_{sor,t} < 0$ (i.e., increased sorption, Fig. 2F and H) for carbonaceous materials. Likewise electrolyte effects on DC-induced PHE desorption from mineral sorbents were observed: increased ionic strength resulted in improved PHE desorption ($\Delta \Gamma_{des,t} > 0$) for silica 40 Å and zeolite NaY (Fig. 2A and C), yet had no apparent impact in AC (($\Delta \Gamma_{des,t} = 0$) Fig. 2E).

4. Discussion

4.1. Effects of the DC field on PHE-sorption and desorption kinetics

Applying a DC field to a suspended solid matrix invokes electrokinetic effects such as EOF, resulting from the motion of mobile counter-ions in the EDL located at the walls of pores and continuous micro-channels (Sinton and Li, 2003; Tallarek et al., 2001). As EOF exerts a dispersive force on PHE molecules while acting at scales relevant for chemical-sorbent interactions, we quantified electrokinetic impacts on the sorption and desorption kinetics of PHE using zeolites, aluminum oxides, silicates, activated carbon and exfoliated graphite. These model geo-sorbents cover a broad range of physicochemical and morphological properties as well as PHE sorption characteristics (Table 1). Suspended sorbents were exposed in a fixed-bed reactor to a weak DC field and differing electrolyte concentrations at constant hydraulic flow. The rates of PHE sorption and desorption were then compared to those in identical DC-free controls. As the EOF velocity ($V_{EOF,r}$; eq. (10)) is directly proportional to $f(\kappa r)$ and ζ; representing two distinct sorbent properties, we presumed that increasing $V_{EOF,r}$ would influence PHE sorption to and PHE desorption from sorption sites beyond mere mass diffusion. Fig. 3C and D correlate observed relative benefits of DC electric fields on PHE desorption ($\Delta \Gamma_{des,t-86h}$) and sorption ($\Delta \Gamma_{sor,t-31h}$) to the calculated $V_{EOF,r}$ (Fig. 3C and D), and the logarithmic surface-normalized PHE partition coefficient (Log K_d^*, Fig. 3A and B) at the end of our experiments (i.e. at 86 h (desorption) and 31 h (sorption). Except for K_d that reflects equilibrium conditions, good correlations ($R^2 = 0.90$ for sorption at 31 h and 0.72 for desorption at 86 h) were detected, supporting the hypothesized effect of EOF on PHE-sorbent interaction (Fig. 3C and D). In order to further evaluate DC-induced kinetic effects we varied the EDL thickness above the sorbents by changing electrolyte concentrations. A shift from 1 to 100 mmol L^{-1} results in a calculated reduction of the EDL thickness from 2.2 to 0.7 nm and an increase of the electroosmotic (EOF) flow velocity in

($V_{EOF,r}$). Simultaneously, a combination of bigger pores and smaller EDL thickness (i.e., increased κr (eq. (10)), promotes an up to five-fold faster $V_{EOF,r}$ and proportional changes $\Delta \Gamma_{des,t}$ and $\Delta \Gamma_{sor,t}$ (Fig. S6) for silica 40 Å, 60 Å, and 100 Å in PB electrolytes of varying ionic strengths. At conditions of low $V_{EOF,r}$ (e.g. when the ionic strength of the electrolyte and/or the zeta potential of the sorbent is low) the DC field-induced impact on PHE-sorbent interactions was low. For the strong PHE sorbents AC and EG, however, no correlation with K_d^*, and $V_{EOF,r}$ was observed (Fig. 3). This suggests that the sorption properties of AC and EG for PHE prevail over the possible EOF effects. As better sorption of PHE to EG in presence of DC was observed, it may be speculated whether EOF may mediate the redistribution of weakly bound PHE within the sorbent (Ai and Jiang, 2012), i.e., translocate PHE molecules from weak to strong sorption sites.

4.2. Sorption thermodynamics and electrokinetic phenomena

In order to further interpret electrokinetic effects on PHE-sorbent interactions, we determined the changes of the Gibbs free energy (ΔG°), the enthalpy (ΔH°) and the entropy (ΔS°) of PHE sorption to all sorbents (Table 2). ΔG° is an indicator for the degree of spontaneity of PHE interaction with sorbents (Liu, 2009). A negative value ($\Delta G^\circ < 0$) thereby refers to a spontaneous reaction. ΔG° of PHE sorption was found to be negative for all sorbents and poorly depended on the ionic strength of the PB electrolyte (Table 2). PHE sorption was exothermic ($\Delta H^\circ < 0$) and accompanied by minor changes of ΔS° (Table 2). This observation confirms earlier work showing that hydrophobic (carbonaceous) and hydrophilic (mineral) surfaces exhibit distinct PAH sorption enthalpies in aqueous solutions (Drost-Hansen, 1978; Huang and Weber, 1997). As the interaction energy of water with mineral surfaces is greater than that of PAH, the water molecules may outcompete PAH molecules in an exothermic sorption processes. According to this assumption, PAH molecules may associate with a ~100 nm thick layer (Drost-Hansen, 1978) of vicinal water rather than directly with the mineral surface (Mader et al., 1997) and, hence, may be subject to significant EOF velocity. The plug-like velocity profile of EOF thereby is likely to exert a dispersing force on PHE molecules above mineral surfaces with a typical electric double layer thickness ranging from 0.65 nm to 6.87 nm for our experimental range of electrolyte concentrations (cf. eq. (13)). According to the model postulated by Huang et al. (Huang and Weber, 1997), the PHE molecules are likely to interact directly with the surface of carbonaceous sorbents (i.e. express clearly negative ΔG°) and hence may require a high EOF kinetic energy for being replaced by water molecules. Based on such reasoning, we tested whether there is an apparent correlation between $V_{EOF,r}$ and the Gibbs free energy for PHE molecule in the vicinity of a sorbent surface (Fig. 4). Such ad hoc correlation is further tempting as the ionic strength of the electrolyte was found to have minor influence on ΔG° yet to promote the intra-pore $V_{EOF,r}$. Figs. 4 and S5 show that $V_{EOF,r}$ as low as 4.4×10^{-7} m s^{-1} result in significant sorption and desorption benefits for sorbents with $\Delta G^\circ > -13.5$ kJ mol^{-1}. Generally, more negative ΔG° and a lower EOF velocity seem to result in an electrokinetic promotion of PHE sorption (cf. warm color area in Fig. 4B) and a reduction of PHE desorption (cf. cold color area in Fig. 4A). By contrast, less negative ΔG° and higher EOF may lead to electrokinetic promotion of PHE desorption (cf. warm color peaks in Fig. 4A) and clearly reduced sorption (cf. cold color peaks in Fig. 4B), respectively.

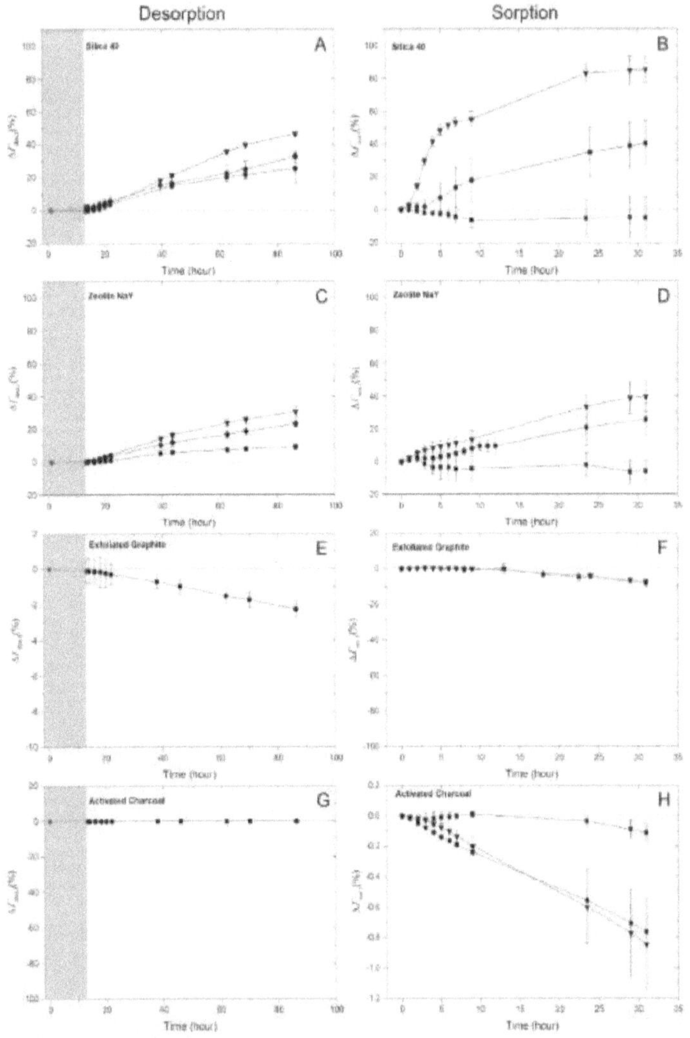

Y. Shan et al. / Chemosphere 242 (2020) 125161

Fig. 2. Effects of PB electrolyte concentrations (1 mmol L^{-1} [squares]; 10 mmol L^{-1} [circles], 50 mmol L^{-1} [diamonds] and 100 mmol L^{-1} [triangles]) on the relative DC-induced influence on PHE desorption ($\Delta\Gamma_{des}$, Fig. 2A, C, E, G) and the DC-induced influence on PHE sorption ($\Delta\Gamma_{sorp}$, Fig. 2B, D, F, H); silica 40 A (Fig. 2A and B), zeolite NaY (Fig. 2C and D), exfoliated graphite (Fig. 2E and F), activated charcoal (Fig. 2G and H). The gray area refers to no electric field periods. Positive and negative values of $\Delta\Gamma_{des}$ refer to increased and reduced desorption in presence of DC, respectively. Positive $\Delta\Gamma_{sorp}$ denotes decreased sorption to the geo-sorbents relative to the control, while negative values refer to increased sorption. Data represent averages and standard deviations of triplicate experiments.

Y. Shan et al. / Chemosphere 242 (2020) 125361

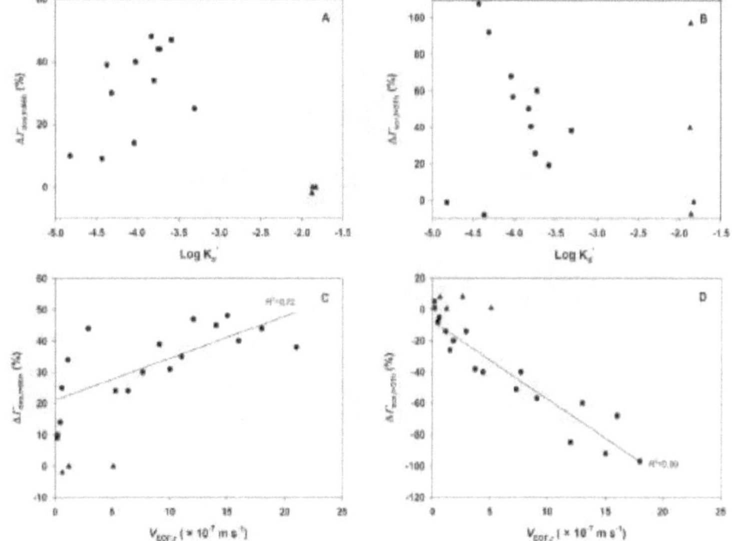

Fig. 3. Effects of the surface normalized partition coefficient K_d' (Fig. 3A and B) and the calculated EOF velocity (Fig. 3C and D) on the relative DC-induced PHE desorption benefits ($\Delta J_{des-mob}$ Fig. 3A and C) and PHE sorption ($\Delta J_{sol-mob}$ Fig. 3B and D). Circles and triangles refer to mineral and carbonaceous (triangles) sorbents, respectively.

Table 2
Table of changes of the Gibbs free energy (ΔG), enthalpy (ΔH) and the entropy (ΔS) of PHE sorbing to different mineral and carbonaceous sorbents at 25 °C in 10 and 100 mmol L^{-1} PB electrolyte solutions.

Sorbents		10 mmol L^{-1}			100 mmol L^{-1}		
		ΔG (kJ mol^{-1})	ΔH (kJ mol^{-1})	ΔS (kJ mol^{-1} K^{-1})	ΔG (kJ mol^{-1})	ΔH (kJ mol^{-1})	ΔS (kJ mol^{-1} K^{-1})
Mineral Sorbents	Silica 40 Å	−11.1	−20.2	−0.03	−9.47	−28.0	−0.06
	Silica 60 Å	−6.9	−20.0	−0.04	−6.85	−25.1	−0.06
	Silica 100 Å	−6.7	−12.3	−0.02	−6.68	−23.7	−0.06
	Zeolite NaY	−13.5	−9.8	0.01	−13.4	−19.0	0.02
	Zeolite 13X	−1.3	−23.8	−0.08	−3.99	−26.3	−0.07
	Al$_2$O$_3$	−4.0	−11.2	0.02	−4.70	−19.1	−0.05
Carbonaceous sorbents	AC	−25.1	−36.3	0.04	−27.6	−20.3	0.01
	BG	−22.6	−32.2	−0.03	−21.3	−54.0	−0.11

4.3. Conclusion: relevance for environmental and biotechnological applications

The transformation of chemicals in natural and man-made systems is often mass transfer-limited as it requires chemicals to be sufficient to initiate chemical or microbial catalysts. In soil for instance, the type, the sorption capacity or the spatial and energetic distribution of the sorption sites may impose serious limitations on the rate of HOC biotransformation. In order to ensure sufficient transformation rates, environmental biotechnology has to ensure and manage transport of chemicals at least over the distances typically separating hotspots of pollution from transforming microbes (Harms and Wick, 2006). This is of special relevance for HOC which are typically associated with solid particles from which they are very slowly released by diffusive transport processes (Johnsen et al., 2005; Semple et al., 2007). Using sorbents of either purely mineral or carbonaceous nature, we here suggest that the application of EOF may be used to control PHE-matrix interactions as a driver for subsequent PHE availability to organisms. In natural systems however, sorbent matrices typically consist of a mixture of mineral and carbonaceous materials and may result in hardly predictable benefits of electrokinetic HOC flushing and biodegradation (Gill et al., 2014; Wick et al., 2007) despite of the absence of apparent negative DC-field effects on soil microbial communities (Wick et al., 2010). Knowledge of the composition of environmental matrices and their chemical, thermodynamic and sorption

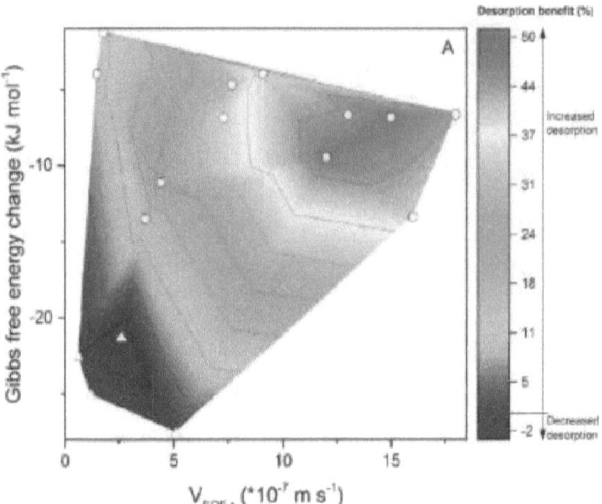

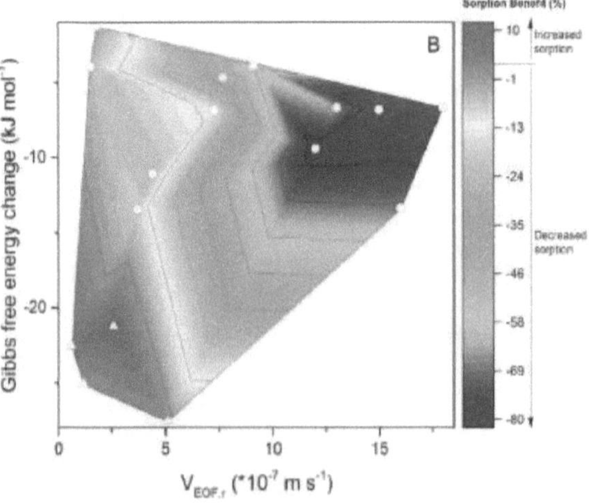

Fig. 4. Apparent effects of the calculated electroosmotic flow velocities ($V_{EOF,r}$) and the Gibbs free energy change (ΔG^0) on the relative DC-induced PHE desorption benefits ($\Delta r_{desorption}$, Fig. 4A) and the relative DC-induced PHE sorption benefits ($\Delta r_{sorption}$, Fig. 4B) using 10 and 100 mmol L^{-1} PB electrolytes.

properties is important for the prediction of electrokinetic effects on HOC-matrix interactions. For example, at situations of low organic carbon content (<1 g kg^{-1}) the mineral phase may dominate and given sufficient electrolyte concentrations DC fields will significantly reduce HOC retention and, hence, increase HOC availability. On the other hand, in activated charcoal treatment of contaminated groundwater or thermal soil remediation technology using activated carbon, DC electric fields may elevate HOC sorption rates, decrease the risk of HOC diffusion, and save the often expensive sorbent materials. Electrokinetic approaches may be further used to kinetically regulate the interaction of sorbates and sorbents in environmental (bio-)technology (Qin et al., 2015a, 2015b; Shan et al., 2018). This kinetic regulation may give rise to future technical applications, which allows regulating sorption processes, for instance in response to fluctuating sorbate concentrations in contaminated water streams, in electro-bioremediation or to avoid unwanted sorption of hydrophobic solutes in technical applications.

CRediT authorship contribution statement

Yongping Shan: Conceptualization, Data curation, Validation, Investigation, Writing - review & editing. **Jinyi Qin:** Conceptualization, Data curation, Formal analysis, Investigation, Writing - review & editing. **Hauke Harms:** Supervision, Validation, Writing - review & editing. **Lukas Y. Wick:** Conceptualization, Data curation, Validation, Funding acquisition, Supervision, Writing - review & editing.

Acknowledgements

This work has been performed in the frame of the Helmholtz Alberta Initiative and contributes to research program topic CITE of the Helmholtz Association. Y. Shan acknowledges financial support by the China Scholarship Council (CSC). The authors wish to thank Kai Uwe Goss (UFZ) and Ulf Roland (UFZ) for helpful discussions as well as Jana Reichenbach, Rita Remer and Birgit Würz for skilled technical help. We are likewise grateful to Ulf Roland for providing the zeolites and for performing the BET analysis of the sorbents.

Appendix A. Supplementary data

Supplementary data to this article can be found online at https://doi.org/10.1016/j.chemosphere.2019.125161.

References

Ahn, S., Werner, D., Karapanagioti, H.K., McGlothlin, D.R., Zare, R.N., Luthy, R.G., 2005. Phenanthrene and pyrene sorption and intraparticle diffusion in polyoxymethylene, coke, and activated carbon. Environ. Sci. Technol. 39, 6516–6526.

Ai, L., Jiang, J., 2012. Removal of methylene blue from aqueous solution with self-assembled cylindrical graphene–carbon nanotube hybrid. Chem. Eng. J. 192, 156–163.

Bansal, V., Kim, K.-H., 2015. Review of PAH contamination in food products and their health hazards. Environ. Int. 84, 26–38.

Cummings, E.B., Griffiths, S.K., Nilson, R.H., Paul, P.H., 2000. Conditions for similitude between the fluid velocity and electric field in electroosmotic flow. Anal. Chem. 72, 2526–2532.

Doke, K.M., Khan, E.M., 2013. Adsorption thermodynamics to clean up wastewater; critical review. Rev. Environ. Sci. Biotechnol. 12, 25–44.

Drost-Hansen, W., 1978. Water at biological interfaces - structural and functional aspects. Phys. Chem. Liq. 7, 243–348.

Duan, L., Naidu, R., 2013. Effect of ionic strength and index cation on the sorption of phenanthrene. Water Air Soil Pollut. 224.

Ghosal, P.S., Gupta, A.K., 2015. An insight into thermodynamics of adsorptive removal of fluoride by calcined Ca–Al–(NO3) layered double hydroxide. RSC Adv. 5, 105889–105900.

Gill, R.T., Harbottle, M.J., Smith, J.W.N., Thornton, S.F., 2014. Electrokinetic-enhanced bioremediation of organic contaminants: a review of processes and environmental applications. Chemosphere 107, 31–42.

Harms, H., Smith, K.E.C., Wick, L.Y., 2017a. Problems of hydrophobicity/bioavailability: an introduction. In: Krell, T. (Ed.), Cellular Ecophysiology of Microbe. Springer International Publishing, Cham, pp. 1–13.

Harms, H., Wick, L.Y., 2006. Dispersing pollutant-degrading bacteria in contaminated soil without touching it. Eng. Life Sci. 6, 252–260.

Harms, H., Wick, L.Y., Smith, K.E.C., 2017b. Matrix-Hydrophobic compound interactions. In: Krell, T. (Ed.), Cellular Ecophysiology of Microbe. Springer International Publishing, Cham, pp. 1–13.

Hassan, I., Mohamedelhassan, E., Yanful, E.K., Yuan, Z.-C., 2015. Sorption of phenanthrene by kaolin and efficacy of hydraulic versus electroosmotic flow to stimulate desorption. J. Environ. Chem. Eng. 3, 2301–2310.

Ho, Y.S., Ng, J.C.Y., McKay, G., 2000. Kinetics of pollutant sorption by biosorbents. Separ. Purif. Methods 29, 189–232.

Hoshyargar, V., Talebi, M., Ashrafizadeh, S.N., Sadeghi, A., 2018. Hydrodynamic dispersion by electroosmotic flow of viscoelastic fluids within a slit microchannel. Microfluid. Nanofluidics 22, 4.

Huang, W., Weber, W.J., 1997. Thermodynamic considerations in the sorption of organic contaminants by soils and sediments. 1. The isosteric heat approach and its application to model inorganic sorbents. Environ. Sci. Technol. 31, 3238–3243.

James, G., Sabatini, D.A., Chiou, C.T., Rutherford, D., Scott, A.C., Karapanagioti, H.K., 2005. Evaluating phenanthrene sorption on various wood chars. Water Res. 39, 549–558.

Jeon, C.-S., Yang, J.-S., Kim, K.-J., Baek, K., 2010. Electrokinetic removal of petroleum hydrocarbon from residual clayey soil following a washing process. Clean. - Soil, Air, Water 38, 189–193.

Johnsen, A.R., Wick, L.Y., Harms, H., 2005. Principles of microbial PAH-degradation in soil. Environ. Pollut. 133, 71–84.

Kim, J.-H., Han, S.-J., Kim, S.-S., Yang, J.-W., 2006. Effect of soil chemical properties on the remediation of phenanthrene-contaminated soil by electrokinetic-Fenton process. Chemosphere 63, 1667–1676.

Kim, J.H., Kim, S.S., Yang, J.W., 2007. Role of stabilizers for treatment of clayey soil contaminated with phenanthrene through electrokinetic-Fenton process—some experimental evidences. Electrochim. Acta 53, 1663–1670.

Kopinke, F.-D., Georgi, A., Goss, K.-U., 2018. Comment on "Mistakes and inconsistencies regarding adsorption of contaminants from aqueous solution: a critical review, published by Tran et al. [Water Research 120, 2017, 88–116]. Water Res. 129, 520–521.

Li, A., Cheung, K.A., Reddy, K.R., 2000. Cosolvent-enhanced electrokinetic remediation of soils contaminated with phenanthrene. J. Environ. Eng. 126, 527–533.

Li, H., Wang, Y., Zheng, H., 2018. Variations of moisture and organics in activated sludge during Fe0/S2O82- conditioning—horizontal electro-dewatering process. Water Res. 129, 83–93.

Liu, Y., 2009. Is the free energy change of adsorption correctly calculated? J. Chem. Eng. Data 54, 1981–1985.

Mader, B.T., Uwe-Goss, K., Eisenreich, S.J., 1997. Sorption of nonionic, hydrophobic organic chemicals to mineral surfaces. Environ. Sci. Technol. 31, 1079–1086.

Morelis, S., van Noort, P.C.M., 2008. Kinetics of phenanthrene desorption from activated carbons to water. Chemosphere 71, 2044–2049.

Moustafa, A.M.A., 2009. Fabrication, Characterization and Oil Spill Remediation Properties of Exfoliated Graphite. MSc. Dissertation. Pennsylvania State University, University park, PA.

Pham, T.D., Shrestha, R.A., Sillanpää, M., 2009. Removal of hexachlorobenzene and phenanthrene from clayey soil by surfactant-and ultrasound-assisted electrokinetics. J. Environ. Eng. 136, 739–742.

Puglisi, E., Cappa, F., Fragoulis, G., Trevisan, M., Del Re, A.A.M., 2007. Bioavailability and degradation of phenanthrene in compost amended soils. Chemosphere 67, 548–556.

Qin, J., Moustafa, A., Harms, H., El-Din, M.G., Wick, L.Y., 2015a. The power of power: electrokinetic control of PAH interactions with exfoliated graphite. J. Hazard Mater. 288, 25–33.

Qin, J., Sun, X., Liu, Y., Berthold, T., Harms, H., Wick, L.Y., 2015b. Electrokinetic control of bacterial deposition and transport. Environ. Sci. Technol. 49, 5663–5671.

Reddy, K.R., 2010. Technical challenges to in-situ remediation of polluted sites. Geotech. Geol. Eng. 28, 211–221.

Reddy, K.R., Maturi, K., Cameselle, C., 2009. Sequential electrokinetic remediation of mixed contaminants in low permeability soils. J. Environ. Eng. 135, 989–998.

https://doi.org/10.1064/ASCE)EE-1042-7670.0000077.

Reddy, K.R., Saichek, R.E., 2004. Enhanced electrokinetic removal of phenanthrene from clay soil by periodic electric potential application. J. Environ. Sci. Health Part A 39, 1189–1212. https://doi.org/10.1081/ESE-120030526.

Rice, C.L., Whitehead, R., 1965. Electrokinetic flow in a narrow cylindrical capillary. J. Phys. Chem. 69, 4017–4024. https://doi.org/10.1021/j100895a062.

Saichek, R.E., Reddy, K.R., 2003. Effect of pH control at the anode for the electrokinetic removal of phenanthrene from kaolin soil. Chemosphere 51, 273–287. https://doi.org/10.1016/S0045-6535(02)00849-4.

Semple, K.T., Doick, K.J., Wick, L.Y., Harms, H., 2007. Microbial interactions with organic contaminants in soil: definitions, processes and measurement. Environ. Pollut. 150, 166–176. https://doi.org/10.1016/j.envpol.2007.07.023.

Shan, Y., Harms, H., Wick, L.Y., 2018. Electric field effects on bacterial deposition and transport in porous media. Environ. Sci. Technol. 52, 14294–14301. https://doi.org/10.1021/acs.est.8b03648.

Shapiro, A.P., Probstein, R.F., 1993. Removal of contaminants from saturated clay by electroosmosis. Environ. Sci. Technol. 27, 283–291. https://doi.org/10.1021/es00039a007.

Sharma, P.K., Rao, K.H., 2001. Adhesion of Paenibacillus polymyxa on chalcopyrite and pyrite: surface thermodynamics and extended DLVO theory. Colloids Surfaces B Biointerfaces 29, 21–38. https://doi.org/10.1016/S0927-7765(02)00186-7.

Shi, L., Harms, H., Wick, L.Y., 2008. Electroosmotic flow stimulates the release of alginate-bound phenanthrene. Environ. Sci. Technol. 42, 2105–2110. https://doi.org/10.1021/es702357ps.

Sinton, D., Li, D., 2003. Electroosmotic velocity profiles in microchannels. Colloids Surf. Physicochem. Eng. Asp. 222, 273–283. https://doi.org/10.1016/S0927-7757(01)00233-4.

Su, Y.-H., Zhu, Y.-G., Sheng, G., Chiou, C.T., 2006. Linear adsorption of nonionic organic compounds from water onto hydrophilic minerals: silica and alumina. Environ. Sci. Technol. 40, 6949–6954. https://doi.org/10.1021/es060669b.

Tallarek, U., Rapp, E., Van As, H., Bayer, E., 2001. Electrokinetics in fixed beds: experimental demonstration of electroosmotic perfusion. Angew. Chem. Int. Ed. 40, 1684–1687. https://doi.org/10.1002/1521-3773(20010504)40:9<1684::AID-ANIE16840>3.0.CO;2-C.

Tran, H.N., You, S.-J., Hosseini-Bandegharaei, A., Chao, H.-P., 2017. Mistakes and inconsistencies regarding adsorption of contaminants from aqueous solutions: a critical review. Water Res. 120, 88–116. https://doi.org/10.1016/j.watres.2017.04.014.

Vallano, P.T., Remcho, V.T., 2000. Modeling interparticle and intraparticle (perfusive) electroosmotic flow in capillary electrochromatography. Anal. Chem. 72, 4255–4265. https://doi.org/10.1021/ac000595b.

Wick, L.Y., Shi, L., Harms, H., 2007. Electro-bioremediation of hydrophobic organic soil-contaminants: a review of fundamental interactions. Electrochim. Acta 52, 3441–3448. https://doi.org/10.1016/j.electacta.2006.03.117.

Wick, L.Y., Buchholz, F., Fetzer, I., Kleinsteuber, S., Harig, C., Miltner, A., Harms, H., Pucci, G.N., 2010. Responses of soil microbial communities to weak electric fields. Sci. Total Environ. 408, 4886–4893. https://doi.org/10.1016/j.scitotenv.2010.06.048.

Zhang, H., Lin, K., Wang, H., Gan, J., 2010. Effect of Pinus radiata derived biochars on soil sorption and desorption of phenanthrene. Environ. Pollut. 158, 2821–2825. https://doi.org/10.1016/j.envpol.2010.06.025.

Zhao, J., Wang, Z., Zhao, Q., Xing, B., 2014. Adsorption of phenanthrene on multilayer graphene as affected by surfactant and exfoliation. Environ. Sci. Technol. 48, 331–339. https://doi.org/10.1021/es4043068c.

5.2 Supporting Information

ELECTROKINETIC EFFECTS ON THE INTERACTION OF PHENANTHERENE WITH GEOSORBENTS

Yongping Shan[1,§], Jinyi Qin[1,§,#], Hauke Harms[1], Lukas Y. Wick[1,*]

[1]Helmholtz Centre for Environmental Research UFZ, Department of Environmental Microbiology, 04318 Leipzig, Germany;

[#]present address: School of Civil Engineering, Chang'an University, Xi'an 710054, China

[*]Corresponding author: Mailing address: Helmholtz Centre for Environmental Research - UFZ, Department of Environmental Microbiology; Permoserstrasse 15; 04318 Leipzig, Germany. phone: +49 341 235 1316, fax: +49 341 235 1351, e-mail: lukas.wick@ufz.de. [§] both authors contributed equally to the manuscript.

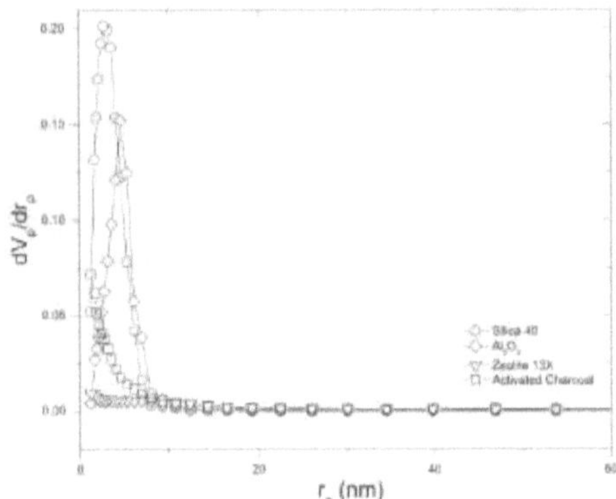

Figure S1. BET result: measured pore size (r_p) distribution relative to pore volume V_p (i.e., dV_p/dr_p) of silica 40, aluminum oxide (Al_2O_3), Zeolite 13X, and activated charcoal.

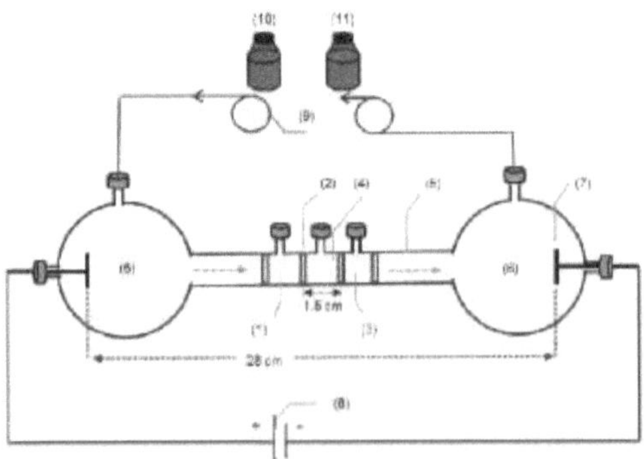

Figure S2. Schematic view of the electrokinetic setup used for sorption and desorption experiments. A reaction chamber packed with the geo-sorbents ($V = 3$ mL; id = 1.5 cm, l = 1.5 cm) (1) is confined by two glass frits (pore size of 160–250 µm; (2)). Chamber (1) is neighbored by two identical chambers (3) with sampling ports (4) for inflow and outflow solute sampling. A glass tube (5) connects the chambers with the electrode compartments ($V \approx 100$ mL; (6)) containing titanium-iridium electrodes (7) as the cathode or anode. The disk-shaped electrodes (7) are welded onto rods which are connected to a DC power pack (8) allowing application of constant electric field strengths. A peristaltic pump (9) supports electrolyte from a light protected reservoir (10) containing dissolved PHE solution in sorption and pure electrolyte in desorption experiments through the anode to the cathode thereby providing a constant hydraulic flow (HF) of 26 mL h^{-1} (11). For minimizing PHE losses all connecting tubes (12) were made of Teflon and the effluent solutions were pumped to a waste container (13). Please note that the setup was used in a vertical position, i.e. with the anode and the cathode compartments (6) being at the top and the bottom of the apparatus, respectively. Please also note that Fig. S1 has been published and described in detail by Qin et al. in a previous study[1]

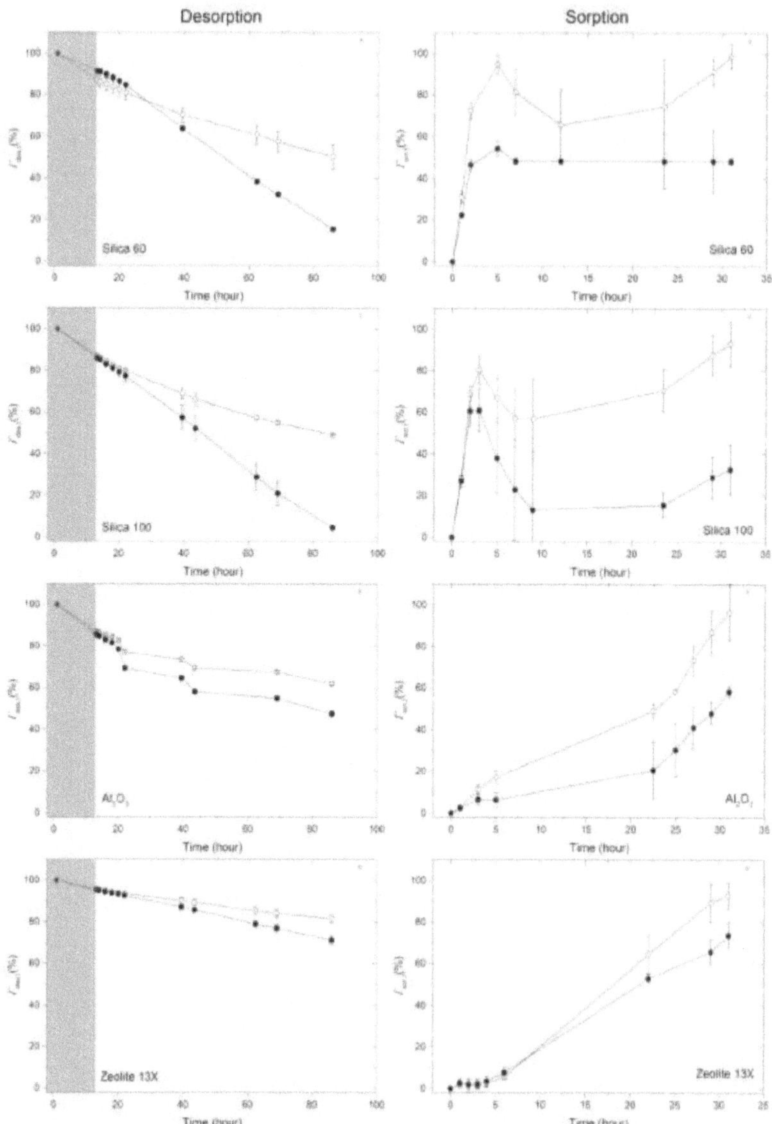

Figure S3. Time-dependent normalized PHE fractions in desorption ($\Gamma_{des,t}$) and sorption ($\Gamma_{sor,t}$) experiments in presence (filled symbols) and absence (open symbols) of a DC electric field using a 10 mmol L^{-1} PB electrolyte: silica 60Å (A & B), silica 100Å (C & D), aluminum oxide (E & F), Zeolite 13X (G & H). Areas with a gray background refer to DC-free periods of the experiments.

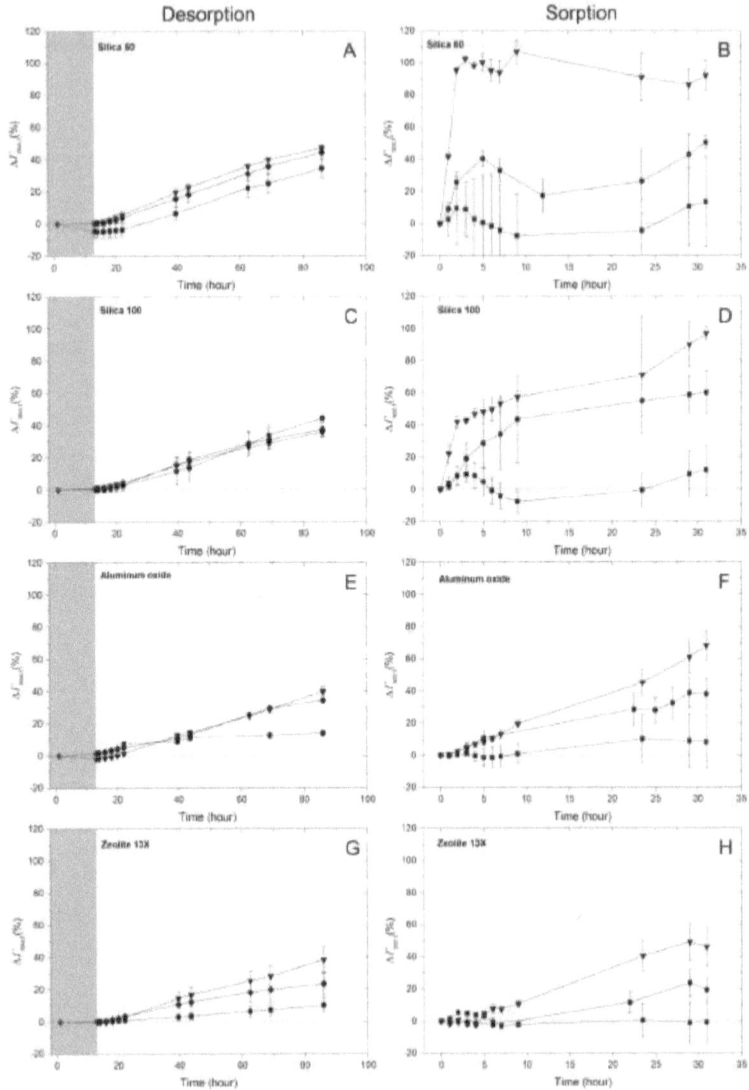

Figure S4. Effects of electrolyte concentrations (1 mmol L^{-1} (squares); 10 mmol L^{-1} (circles), 50 mmol L^{-1} (diamonds) and 100 mmol L^{-1} (triangles)) on the relative DC-induced influence on PHE desorption ($\Delta\Gamma_{des,t}$: A, C, E, G) and PHE sorption ($\Delta\Gamma_{sor,t}$: B, D, F, H): silica 60Å (A & B), silica 100Å (C & D), aluminum oxide (E & F), zeolite 13X (G & H). The grey area refers to no electric field periods. Positive and negative values of $\Delta\Gamma_{des,t}$ refer to increased and reduced desorption in the presence of DC, respectively. Positive $\Delta\Gamma_{sor,t}$ denotes decreased sorption to the geo-sorbents relative to the control, while negative values refer to increased sorption.

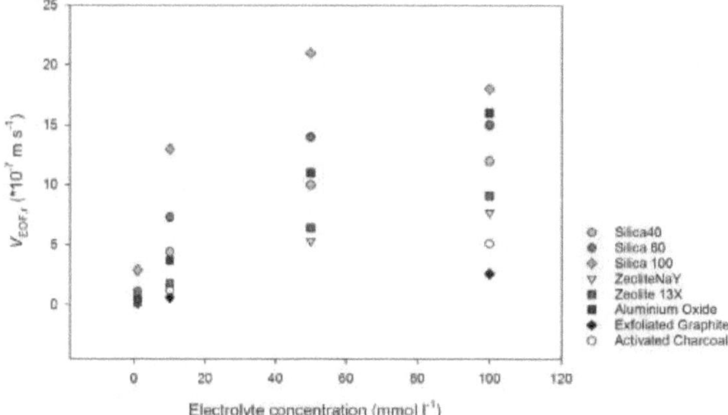

Figure S5. Calculated electroosmotic flow velocities (cf. eq. 10) in intra-particle pores ($V_{EOF,i}$) of eight model geo-sorbents at different potassium phosphate buffer electrolyte concentrations assuming corresponding average pore sizes as given in Table 1.

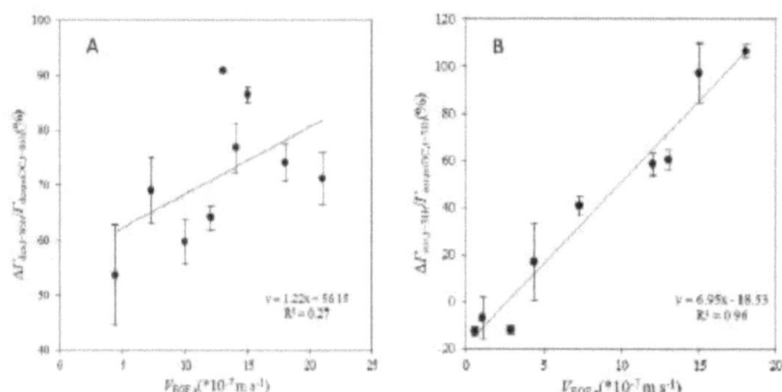

Figure S6. Normalized DC benefit on desorption (Fig. S6A) and sorption (Fig. S6B) vs. the calculated electroosmotic flow velocities (cf. eq 10) in intra-particle pores ($V_{EOF,i}$) of silica 40Å, silica 60Å, and silica 100Å at different potassium phosphate buffer electrolyte concentrations (10, 50, and 100 mM for desorption; 1, 10, and 100 mM for sorption) assuming corresponding average pore sizes as given in Table 1.

REFERENCE

(1) Qin, J.; Sun, X.; Liu, Y.; Berthold, T.; Harms, H.; Wick, L. Y. Electrokinetic Control of Bacterial Deposition and Transport. Environ. Sci. Technol. 2015, 49 (9), 5663–5671. https://doi.org/10.1021/es506245y.

6. Discussion and Outlook

The principles of electrokinetic effects on bacterial deposition in porous media, bacterial deposition on planar surfaces, and contaminant release from a matrix were investigated separately in the three work packages. Investigations found that electrokinetics can exert strong effects on both bacteria deposition and contaminant release under various matrix and electrolyte conditions. The driving factors of electrokinetic effects were further analyzed by interlinking the observations from experiments and theoretical quantified electrokinetic strengths. Besides revealing the principles of electrokinetic effects, these approaches open the possibilities of predicting the improvement of contaminant availability in various matrices and environmental conditions.

6.1 Driving Factors of Electrokinetic Effects on Bacterial Deposition

6.1.1 Establishing an Approach to Predict Electrokinetic Effects on Bacterial Deposition

Based on previous research[26] that interlinked enhanced *P. fluorescens* LP6a transport with electroosmotic shear force (F_{EOF}), we challenged the F_{EOF}-effects model by four bacterial strains differing in physicochemical properties. Besides confirmed transport-enhancing effects for two strains LP6a and S3, we simultaneously found that DC electric fields significantly promoted the deposition efficiencies of *P. putida* KT2440 and *R. opacus* X9. Further investigations on the driving factors of deposition, we found that in addition to the DLVO force (F_{DLVO}), hydraulic shear force (F_{HF}), and electroosmotic shear force (F_{EOF}), there is another unneglectable driving force on bacteria deposition: the electrophoretic drag force (F_{EP}). After inducing $F_{EP,}$ the relative changes of electrokinetic-induced net forces were highly correlated with the relative changes of the deposition efficiency. The relative changes of net forces originate from the electroosmotic shear force and electrophoretic force, they have the same magnitude but point to opposite directions when both bacteria and matrix surface carry negative charges. For all bacterial strains, at both initial and final stages of the breakthrough curves, the relative strengths of these two forces ($|F_{EOF}|$ and $|F_{EP}|$) determined the electrokinetic effects. At the conditions $|F_{EOF}| > |F_{EP}|$ increased deposition efficiency was detected, while at conditions $|F_{EOF}| < |F_{EP}|$ decreased deposition efficiency was detected. These findings hence support that electrokinetic shear and drag forces are driving factors of electrokinetic-effects on bacterial deposition. In high ionic strength, the driving factor $|F_{EOF}|/|F_{EP}|$ can be approximately simplified to the function of the zeta potentials of the bacteria and the matrix surface by 1.29 ζ_C/ζ_{bac}, which can be further used for quick predictions of electrokinetic effects. At 1.29 $\zeta_C/\zeta_{bac} < 1$ reduced deposition is to be predicted and at 1.29 $\zeta_C/\zeta_{bac} > 1$ increased deposition is to be predicted. The established approach provides the possibility to manipulate and to predict the electrokinetic effects on bacterial deposition.

6.1.2 Evidencing the Electrokinetic Bacterial Deposition Approach with QCM-D Monitoring

In the porous media, an approach has been established to describe the principles and to predict the electrokinetic effects on bacterial deposition. However, direct measured evidence of the strength/rigidity of the bacterial attachment was not available by column observations. Therefore, we further challenged to prove the electrokinetic effects on bacterial deposition by QCM-D monitoring and microscopy approaches. This real-time method measures the rigidity of bacterial attachment by QCM-D monitoring and microscope counting under various electric field conditions at a single bacterium weight level [100] at several nanometers above a planar surface. The electrokinetic effects on the deposition of two bacterial strains LP6a and KT2440 in QCM-D system in wider electrolyte concentration range were investigated. Investigations found that, for both strains LP6a and KT2440, electrokinetics significantly changed bacterial deposition in electrolyte concentrations from 10 to 100 mM. We hence correlated electrokinetic-induced deposition effects to F_{net} shifts; i.e. the QCM-D measured attachment rigidity ($\Delta f_5/\Delta D_5$) and the microscope observed cell density (η_c), to F_{net} acting on a bacterium at the secondary minimum above the sensor surface. We found that the rigidity, the cell density fit well to the previous established approach of electrokinetic effects on bacterial deposition. We found good apparent correlation between the normalized microscopy cell density η_c and QCM-D derived rigidity at all electric field strengths and electrolyte concentrations tested. Increasing attachment rigidity was mirrored by higher cell counts, while decreasing attachment rigidity resulted in lower cell counts.

This highlights QCM-D as a useful approach to assess and predict the influence of DC electric fields on bacterial deposition in combination with the established approach of electrokinetic effects on bacterial deposition: At $F_{net,DC} > F_{net,no\ DC}$ increased attachment rigidity and η_c and at $F_{net,DC} < F_{net,noDC}$ lowered attachment rigidity and η_c as compared to DC free controls were observed. As F_{EOF} and F_{EP} are of opposite sign in our experimental system, their relative strength is a driver of $F_{net,DC}$ and, thus, of observed electrokinetic effects on bacterial deposition. If $|F_{EOF}| > |F_{EP}|$, electrokinetics promote attachment rigidity and η_c and *vice versa*, respectively[44]. The $|F_{EOF}| / |F_{EP}|$ thus was a good predictor for bacterial electrokinetic effects on cell attachment rigidity and bacterial deposition at all conditions tested. The heat maps in Figs. 4 & S8 in Chapter 3 visualized the effects of $|F_{EOF}|$ and $|F_{EP}|$ on normalized DC-induced rigidity and η_c changes. They reveal the importance of the $|F_{EP}|$ for cell deposition at given $|F_{EOF}|$ independent of the strain, electrolyte strength or the electric field applied. The high degree of convergence of rigidity and η_c changes further proposes that QCM-D is a good and fast tool for real-time analysis of electrokinetic deposition. Simultaneously, with the increment of all electrolyte concentrations from 10 to 100 mM, the electrokinetic effects on LP6a varied from transport-enhancing to much stronger transport-enhancing, while effects on KT2440 varied from transport-enhancing to deposition-enhancing. QCM-D research evidenced the electrokinetic effects and provided the possibility of controlling the electrokinetic effects by varying the DC field conditions (e.g., steering electrokinetic effects of strain KT2440).

6.2 Driving Factors of Electrokinetic Effects on the Interactions between Matrices and Contaminant

In the ecosystems, the type, the sorption capacity, or the spatial and energetic distribution of the sorption sites may impose serious limitations on the rate of PAH biotransformation. In order to ensure sufficient transformation rates, environmental biotechnology has to manage the transport of contaminants at least over the distances typically separating hotspots of pollution from transforming microbes [101]. This is of special relevance for PAH which are typically associated with solid particles from which they are very slowly released by diffusive transport processes [7,102]. Using sorbents of either purely mineral or carbonaceous nature, we here suggest that the application of electrokinetics may be used to control PHE-matrix interactions as a driver for subsequent PHE availability to microbes. Applying a DC field to a solid matrix invokes electroosmotic flow, resulting from the relative motion of counter-ions adsorbed on the inner and outer surface of pores and continuous micro-channels [43,103]. As EOF exerts dispersive effects on PHE molecules, we quantified electrokinetic impacts on the sorption and desorption kinetics of PHE using zeolites, aluminum oxides, silicates, activated carbon and exfoliated graphite. The rates of PHE sorption and desorption were then compared to those in identical DC-free controls. Our research correlated observed relative effects of DC fields on PHE sorption and desorption to the calculated $V_{EOF,r}$ in various sorbents and intra-pore size, good correlations were detected. These results support the hypothesized effect of EOF on PHE-sorbent interaction. In order to further evaluate DC-induced kinetic effects, we varied the electrical double layer thickness by changing electrolyte concentrations. A shift of electrolyte concentration from 1 to 100 mmol L^{-1} results in a reduction of the EDL thickness and an increase of the electroosmotic flow velocity in microchannels ($V_{EOF,r}$). Simultaneously, a combination of bigger pores and smaller EDL thickness promotes an up to fivefold-faster $V_{EOF,r}$ and proportional changes of electrokinetic effects on sorption and desorption for silica sorbents in varying electrolyte concentrations. At conditions of low $V_{EOF,r}$, the DC field-induced impact on PHE-sorbent interactions was low. For the strong PHE sorbents activated carbon (AC) and exfoliated graphite (EG), however, no correlation with $V_{EOF,r}$ was observed. This suggests that the sorption properties of AC and EG for PHE prevail over the possible EOF effects. As better sorption of PHE to EG in the presence of DC was observed, it may be speculated whether EOF may mediate the redistribution of weakly bound PHE within the sorbent [104], i.e., translocate PHE molecules from weak to strong sorption sites.

In order to further interpret the DC-effects on PHE-sorbent interactions, we determined the thermodynamic parameters such as the Gibbs free energy changes (ΔG°), the enthalpy change (ΔH°), and the entropy change (ΔS°) of PHE sorption to sorbents. ΔG° as an indicator of the degree of the spontaneity of PHE interaction with sorbents[105] was found to be negative and poorly dependent on the ionic strength of electrolyte. PHE sorption was exothermic ($\Delta H^\circ < 0$) and accompanied by minor changes of ΔS°. This observation confirms earlier work showing that hydrophobic (carbonaceous) and hydrophilic (mineral) surfaces exhibit distinct PAH sorption enthalpies in aqueous solutions [92,106]. As the interaction energy of water with mineral surfaces is greater than that of PAH, the water molecules may outcompete PAH molecules in an exothermic sorption process. According to this assumption, PAH molecules may associate with a ~ 100 nm thick layer [106] of vicinal water rather than directly with the mineral surface [107] and, hence, may be subject to significant EOF velocity. The plug-like velocity profile of EOF thereby is likely to exert a dispersing force on PHE molecules above mineral surfaces with a typical electric double layer thickness ranging from 0.65 nm to 6.87 nm for our experimental range of electrolyte concentrations. According to the model postulated by Huang et al.[92], the PHE molecules are likely to interact directly with the surface of carbonaceous sorbents and hence may require high ΔG° for their replacement by water molecules. Based on such reasoning, we tested whether there is an apparent correlation between $V_{EOF,r}$ and the Gibbs free energy for PHE molecule in the vicinity of a sorbent surface. Such correlation further tempting as the ionic strength of the electrolyte was found to have minor influences on ΔG° yet to promote the intra-pore $V_{EOF,r}$. $V_{EOF,r}$ as low as 4.4×10^{-7} m s^{-1} results in significant apparent sorption and desorption benefits for sorbents with $\Delta G^\circ > -13.5$ kJ mol^{-1}. Generally, higher ΔG° and a lower EOF velocity seem to result in an electrokinetic promotion of PHE sorption and a reduction of PHE desorption. By contrast, lower ΔG° and higher EOF may lead to an electrokinetic promotion of PHE desorption and clearly reduced sorption, respectively.

6.3 Relevance for Environmental Application

Using the established approaches interlinking the electrokinetic forces with DLVO interactions energies and sorption energy ΔG°, we were able to estimate electrokinetic effects on bacterial deposition and contaminant release, respectively. The electrokinetic control bacterial deposition approach can be used to control and predict electrokinetic effects on bacteria deposition efficiency in both porous and planar systems; and the electrokinetic control contaminant release approach can be used to control and predict the electrokinetic effects on contaminant sorption and desorption. Based on these findings, we would like to seek the possibility of using one or both of the approaches to help the applications in the ecosystem.

Our approach on electrokinetic control bacterial deposition can be used to predict bacterial deposition rate in DC fields, to find the possibility of applying electrokinetic to disperse functional bacteria in ecosystems to enhance availability or preventing bacteria pollution of the drinking water system. As described in chapter 6.1, in high ionic strength, the driving factors of electrokinetic effects can be simplified to the zeta potential ratio of the matrix and bacteria (i.e., ζ_C/ζ_{bac}). Investigations have found that in the natural soil system where typical zeta potential distribution ranges of bacteria (-5 to -48 mV[108,109]) and matrices (0 to -54 mV[110–112]) are relatively wide, the two different effects of electric fields exist at the same time regarding the ζ_C/ζ_{bac} distribution. For the situations 1.29 $\zeta_C/\zeta_{bac} > 1$ (i.e. $|F_{EOF}| > |F_{EP}|$), DC fields enhance the deposition of bacterial in porous matrices, however, the strong F_{EOF} may enhance the desorption and migration of contaminants, and thus may also bridge the physical distance between bacterium and contaminants to further enhance bioremediation. On the other hand, at 1.29 $\zeta_C/\zeta_{bac} < 1$ (i.e. $|F_{EOF}| < |F_{EP}|$), DC fields may enhance the transport of bacteria through porous media to reach contaminants adsorbed on matrices, and enhance bioremediation. Increasing the surface charge of the matrices (ζ_C) results in increasing net force, therefore supports the deposition of bacteria and may promote desired biofilm formation. While decreasing ζ_C / ζ_{bac} results in the decreasing net force, therefore reduces bacterial deposition and, hence, transport bacterial cells deep into the porous media. In addition, based on the QCM-D evidenced electrokinetic effects in various ionic strength, there is a possibility of steering the DC-effects on bacterial desposition by varying the DC field conditions. In electrokinetically-managed natural and man-made ecosystems the electrokinetic control bacterial deposition approach and QCM-D real-time measurement method hence allow for better control of microbial deposition and transport to achieve better bioavailability.

Our approach to estimate electrokinetic control contaminant release can be used to predict the contaminant releasing in DC fields, to find the possibility of improving contaminant releasing in soil system to enhance availability or to prevent pollutants invading into clean environment systems. Knowledge of the composition of environmental matrices and their chemical, thermodynamic and sorption properties is important for the prediction of electrokinetic effects on PAH-matrix interactions. For example, at situations of low organic carbon content (< 1 g kg^{-1}) the mineral phase may dominate and given sufficient electrolyte concentrations, DC fields will significantly reduce PAH retention and, hence, increase PAH availability. On the other hand, in activated charcoal treatment of contaminated groundwater or thermal soil remediation technology using activated carbon, DC electric fields may elevate PAH sorption rates, decrease the risk of PAH diffusion, and save the often expensive sorbent materials. Electrokinetic approaches may be further used to kinetically regulate the interaction of sorbates and sorbents in environmental (bio-)technology [24,26,44]. This kinetic regulation may give rise to future technical applications, which allows regulating sorption processes, for instance in response to fluctuating sorbate concentrations in contaminated water streams, in electro-bioremediation or to avoid unwanted sorption of hydrophobic solutes in technical applications.

Overall considering both electrokinetic effect approaches, we found that electrokinetics allows for very wide application opportunities to improve bioavailability in various ecosystems as described by Fig. 8.

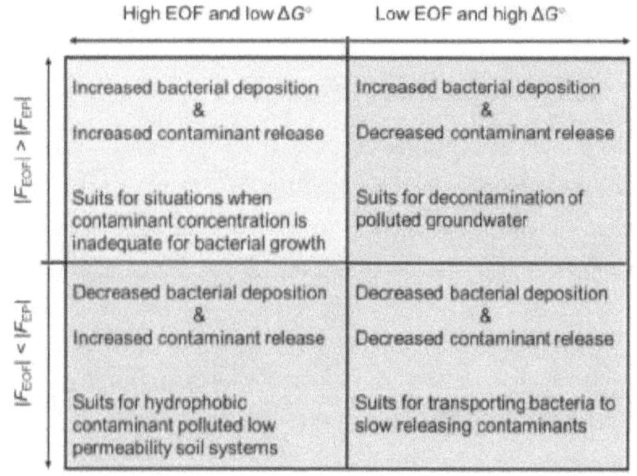

	High EOF and low $\Delta G°$	Low EOF and high $\Delta G°$
$|F_{EOF}| > |F_{EP}|$	Increased bacterial deposition & Increased contaminant release Suits for situations when contaminant concentration is inadequate for bacterial growth	Increased bacterial deposition & Decreased contaminant release Suits for decontamination of polluted groundwater
$|F_{EOF}| < |F_{EP}|$	Decreased bacterial deposition & Increased contaminant release Suits for hydrophobic contaminant polluted low permeability soil systems	Decreased bacterial deposition & Decreased contaminant release Suits for transporting bacteria to slow releasing contaminants

Figure. 8 Schematic of strategies of applying electrokinetics in various conditions

In $|F_{EOF}| > |F_{EP}|$, high EOF, and low $\Delta G°$ conditions (cf. the orange area in Fig. 8), electrokinetics results in increased bacterial deposition and increased contaminant release. This suits for improving the bioavailability in the contaminated sites when the concentration of contaminants is not adequate for bacterial growth (for instance, Rein et al. found that concentrations of < 5-10 nM of polycyclic aromatic hydrocarbons (PAH) did not meet the maintenance requirements of the degrader population). In $|F_{EOF}| > |F_{EP}|$, low EOF, and high $\Delta G°$ conditions (cf. the red area in Fig. 8), electrokinetics results in increased bacterial deposition and decreased contaminant release. This suits for the decontamination of groundwater pollutants by enhancing bacterial deposition and increasing contaminant concentration for biofilm growth. In $|F_{EOF}| < |F_{EP}|$, high EOF, and low $\Delta G°$ conditions (cf. the green area in Fig. 8), electrokinetics results in decreased bacterial deposition and increased contaminant release. This suits for enhancing bioavailability in low permeability ecosystems (for instance PAH contaminated oil sand sites) by enhancing bacterial mobility and contaminant release/transport to degraders. In this case, electrokinetics can also be applied to disperse high surface charged specific degraders into the contaminated sites. In $|F_{EOF}| < |F_{EP}|$, low EOF, and high $\Delta G°$ conditions (cf. the blue area in Fig. 8), electrokinetics results in decreased bacterial deposition and decreased contaminant release. This suits for improving availability by transporting bacterial degraders to habitats where the contaminants are released at a very low rate.

In the meanwhile, based on the driving factors of electrokinetic, there are possibilities to steer the electrokinetic effects to fit the need of specified ecosystems. For instance, increasing the electrolyte concentration allows for changing the relative strength of $|F_{EOF}|$ and $|F_{EP}|$, therefore it may steer the electrokinetic effects from decreasing deposition to increasing deposition. On the other hand, adding surfactants may decrease the sorption Gibbs free energy $\Delta G°$, therefore it may reverse the relative strength of EOF and $\Delta G°$ to steer the electrokinetic effects from decreasing release to increasing release.

6.4 Outlook

Research of this book established two electrokinetic effect approaches to predict the DC effects on bacterial deposition and contaminant release. Based on these approaches we can foresee that there are plenty of application opportunities in ecosystems. We hence propose several research interests following the current knowledge of electrokinetic effects:

Firstly, electrokinetic effects on the biodegradation efficiency. Currently we investigated the electrokinetic effects on the driving factors of bioavailability, i.e., the first part of bioavailability (cf. Fig. 1), it is interesting further to investigate electrokinetic effects on the overall degradation efficiency when the availability of contaminant to degraders is increased.

Secondly, the two electrokinetic effect approaches also give the driving factors that may change or even steer the DC-effects to adjust the electrokinetic effect to fit for the needs of specified ecosystems. Therefore, it will be interesting to seek for more ecologically-friendly and operation-practical methods in controlling the DC-effects to the wanted direction. For instance, the surface charge of degraders may be modified by varying the medium as described by Simoni et al.[113,114], this allows for changes of the relative strength of F_{EOF} and F_{EP} and therefore may change the DC-effects on bacterial deposition efficiency. On the other hand, adding surfactants allows for changes of the relative strength of EOF and the $\Delta G°$ and therefore may change the DC-effects on contaminant release.

In addition, natural ecosystems are complex, as more organisms exist besides bacteria such as worms, plants, fungi, etc. It will be very interesting and required to investigate the electrokinetic effects on these living organisms to define the appropriate application conditions of DC fields.

7. References

(1)   Harmsen, J. Measuring Bioavailability: From a Scientific Approach to Standard Methods. *J. Environ. Qual.* **2007**, *36* (5), 1420–1428. https://doi.org/10.2134/jeq2006.0492.

(2)   Harms, H.; Smith, K. E. C.; Wick, L. Y. Problems of Hydrophobicity/Bioavailability: An Introduction. In *Cellular Ecophysiology of Microbe*; Krell, T., Ed.; Springer International Publishing: Cham, 2017; pp 1–13. https://doi.org/10.1007/978-3-319-20796-4_38-1.

(3)   Bosma, T. N. P.; Middeldorp, P. J. M.; Schraa, G.; Zehnder, A. J. B. Mass Transfer Limitation of Biotransformation: Quantifying Bioavailability. *Environ. Sci. Technol.* **1997**, *31* (1), 248–252. https://doi.org/10.1021/es960383u.

(4)   Danielopol, D. L.; Pospisil, P.; Rouch, R. Biodiversity in Groundwater: A Large-Scale View. *Trends Ecol. Evol.* **2000**, *15* (6), 223–224. https://doi.org/10.1016/S0169-5347(00)01868-1.

(5)   König, S.; Worrich, A.; Banitz, T.; Harms, H.; Kästner, M.; Miltner, A.; Wick, L. Y.; Frank, K.; Thullner, M.; Centler, F. Functional Resistance to Recurrent Spatially Heterogeneous Disturbances Is Facilitated by Increased Activity of Surviving Bacteria in a Virtual Ecosystem. *Front. Microbiol.* **2018**, *9*, 734.

(6)   Worrich, A.; Koenig, S.; Miltner, A.; Banitz, T.; Centler, F.; Frank, K.; Thullner, M.; Harms, H.; Kaestner, M.; Wick, L. Y. Mycelium-Like Networks Increase Bacterial Dispersal, Growth, and Biodegradation in a Model Ecosystem at Various Water Potentials. *Appl. Environ. Microbiol.* **2016**, *82* (10), 2902–2908. https://doi.org/10.1128/AEM.03901-15.

(7)   Johnsen, A. R.; Wick, L. Y.; Harms, H. Principles of Microbial PAH-Degradation in Soil. *Environ. Pollut.* **2005**, *133* (1), 71–84. https://doi.org/10.1016/j.envpol.2004.04.015.

(8)   Worrich, A.; König, S.; Banitz, T.; Centler, F.; Frank, K.; Thullner, M.; Harms, H.; Miltner, A.; Wick, L. Y.; Kästner, M. Bacterial Dispersal Promotes Biodegradation in Heterogeneous Systems Exposed to Osmotic Stress. *Front. Microbiol.* **2016**, *7*, art. 1214. https://doi.org/10.3389/fmicb.2016.01214.

(9)   Wick, L. Y.; Harms, H.; Smith, K. E. C. Microorganism-Hydrophobic Compound Interactions. In *Cellular Ecophysiology of Microbe*; Krell, T., Ed.; Springer International Publishing: Cham, 2017; pp 1–15. https://doi.org/10.1007/978-3-319-20796-4_40-1.

(10)   Rein, A.; Adam, I. K. U.; Miltner, A.; Brumme, K.; Kästner, M.; Trapp, S. Impact of Bacterial Activity on Turnover of Insoluble Hydrophobic Substrates (Phenanthrene and Pyrene)—Model Simulations for Prediction of Bioremediation Success. *J. Hazard. Mater.* **2016**, *306*, 105–114. https://doi.org/10.1016/j.jhazmat.2015.12.005.

(11)   Wick, L. Y.; Shi, L.; Harms, H. Electro-Bioremediation of Hydrophobic Organic Soil-Contaminants: A Review of Fundamental Interactions. *Electrochimica Acta* **2007**, *52* (10), 3441–3448. https://doi.org/10.1016/j.electacta.2006.03.117.

(12)   Shi, L.; Harms, H.; Wick, L. Y. Electroosmotic Flow Stimulates the Release of Alginate-Bound Phenanthrene. *Environ. Sci. Technol.* **2008**, *42* (6), 2105–2110. https://doi.org/10.1021/es702357p.

(13)   Shi, L.; Mueller, S.; Harms, H.; Wick, L. Y. Effect of Electrokinetic Transport on the Vulnerability of PAH-Degrading Bacteria in a Model Aquifer. *Environ. Geochem. Health* **2008**, *30* (2), 177–182. https://doi.org/10.1007/s10653-008-9146-0.

(14)   Bridgett, M. J.; Davies, M. C.; Denyer, S. P. Control of Staphylococcal Adhesion to Polystyrene Surfaces by Polymer Surface Modification with Surfactants. *Biomaterials* **1992**, *13* (7), 411–416. https://doi.org/10.1016/0142-9612(92)90159-L.

(15)   Multiple Interactions in Riverine Biofilms ? Surfactant Adsorption, Bacterial Attachment and Biodegradation. *Water Sci. Technol.* **1995**, *31* (1). https://doi.org/10.1016/0273-1223(95)00155-G.

(16)   Paul, J. H.; Jeffrey, W. H. The Effect of Surfactants on the Attachment of Estuarine and Marine Bacteria to Surfaces. *Can. J. Microbiol.* **1985**, *31* (3), 224–228. https://doi.org/10.1139/m85-043.

(17)   Marco Papini Petrangeli; Mauro Majone; Firoozeh Arjmand; Daniele Silvestri; Marco Sagliaschi; Salvatore Sucato; Eduard Alesi. First Pilot Test on Integration of Gcw (Groundwater Circulation Well) with Ena (Enhanced Natural Attenuation) for Chlorinated

Solvents Source Remediation. *Chem. Eng. Trans.* **2016**, *49*, 91–96. https://doi.org/10.3303/CET1649016.

(18) Johnson, R. L.; Simon, M. A. Evaluation of Groundwater Flow Patterns around a Dual-Screened Groundwater Circulation Well. *J. Contam. Hydrol.* **2007**, *93* (1–4), 188–202. https://doi.org/10.1016/j.jconhyd.2007.02.003.

(19) Alesi, E. J.; Leins, C. H. Multifunctional Well Implemented for In Situ Remediation of Hydrocarbon Contaminations in Soil and Groundwater. In *Contaminated Soil '95*; Brink, W. J., Bosman, R., Arendt, F., Eds.; Springer Netherlands: Dordrecht, 1995; pp 1135–1136. https://doi.org/10.1007/978-94-011-0421-0_41.

(20) Trellu, C.; Mousset, E.; Pechaud, Y.; Huguenot, D.; van Hullebusch, E. D.; Esposito, G.; Oturan, M. A. Removal of Hydrophobic Organic Pollutants from Soil Washing/Flushing Solutions: A Critical Review. *J. Hazard. Mater.* **2016**, *306*, 149–174. https://doi.org/10.1016/j.jhazmat.2015.12.008.

(21) Gill, R. T.; Harbottle, M. J.; Smith, J. W. N.; Thornton, S. F. Electrokinetic-Enhanced Bioremediation of Organic Contaminants: A Review of Processes and Environmental Applications. *Chemosphere* **2014**, *107*, 31–42. https://doi.org/10.1016/j.chemosphere.2014.03.019.

(22) Annamalai, S.; Sundaram, M. Electro-Bioremediation: An Advanced Remediation Technology for the Treatment and Management of Contaminated Soil. In *Bioremediation of Industrial Waste for Environmental Safety*; Bharagava, R. N., Saxena, G., Eds.; Springer Singapore: Singapore, 2020; pp 183–214. https://doi.org/10.1007/978-981-13-3426-9_8.

(23) Beretta, G.; Mastorgio, A. F.; Pedrali, L.; Saponaro, S.; Sezenna, E. The Effects of Electric, Magnetic and Electromagnetic Fields on Microorganisms in the Perspective of Bioremediation. *Rev. Environ. Sci. Bio-Technol.* **2019**, *18* (1), 29–75. https://doi.org/10.1007/s11157-018-09491-9.

(24) Qin, J.; Moustafa, A.; Harms, H.; El-Din, M. G.; Wick, L. Y. The Power of Power: Electrokinetic Control of PAH Interactions with Exfoliated Graphite. *J. Hazard. Mater.* **2015**, *288*, 25–33. https://doi.org/10.1016/j.jhazmat.2015.02.008.

(25) Acar, Y.; Alshawabkeh, A. Principles of Electrokinetic Remediation. *Environ. Sci. Technol.* **1993**, *27* (13), 2638-+. https://doi.org/10.1021/es00049a002.

(26) Qin, J.; Sun, X.; Liu, Y.; Berthold, T.; Harms, H.; Wick, L. Y. Electrokinetic Control of Bacterial Deposition and Transport. *Environ. Sci. Technol.* **2015**, *49* (9), 5663–5671. https://doi.org/10.1021/es506245y.

(27) Shi, L.; Müller, S.; Loffhagen, N.; Harms, H.; Wick, L. Y. Activity and Viability of Polycyclic Aromatic Hydrocarbon - degrading Sphingomonas Sp. LB126 in a DC - electrical Field Typical for Electrobioremediation Measures. *Microb. Biotechnol.* **2008**, *1* (1), 53–61.

(28) Hu, Y.; Werner, C.; Li, D. Electrokinetic Transport through Rough Microchannels. *Anal. Chem.* **2003**, *75* (21), 5747–5758. https://doi.org/10.1021/ac0347157.

(29) Pennathur, S.; Santiago, J. G. Electrokinetic Transport in Nanochannels. 1. Theory. *Anal. Chem.* **2005**, *77* (21), 6772–6781. https://doi.org/10.1021/ac050835y.

(30) Masliyah, J. H.; Bhattacharjee, S. *Electrokinetic and Colloid Transport Phenomena*; John Wiley & Sons: New Jersey, 2006.

(31) Kuo, C.-C.; Papadopoulos, K. D. Electrokinetic Movement of Settled Spherical Particles in Fine Capillaries. *Environ. Sci. Technol.* **1996**, *30* (4), 1176–1179. https://doi.org/10.1021/es950413d.

(32) *Particle Deposition and Aggregation: Measurement, Modelling and Simulation*; Elimelech, M., Ed.; Colloid and surface engineering series; Butterworth-Heinemann: Oxford, 1998.

(33) Hunter, R. J. *Zeta Potential in Colloid Science: Principles and Applications*, 3. print.; Colloid science; Academic Pr: London, 1988.

(34) *Interfacial Electrokinetics and Electrophoresis*; Delgado, Á. V., Ed.; Surfactant science series; Marcel Dekker, Inc: New York, 2002.

(35) Clogston, J. D.; Patri, A. K. Zeta Potential Measurement. In *Characterization of Nanoparticles Intended for Drug Delivery*; McNeil, S. E., Ed.; Humana Press: Totowa, NJ, 2011; Vol. 697, pp 63–70. https://doi.org/10.1007/978-1-60327-198-1_6.

(36)   Xu, R. Progress in Nanoparticles Characterization: Sizing and Zeta Potential Measurement. *Particuology* **2008**, *6* (2), 112–115. https://doi.org/10.1016/j.partic.2007.12.002.

(37)   Sze, A.; Erickson, D.; Ren, L.; Li, D. Zeta-Potential Measurement Using the Smoluchowski Equation and the Slope of the Current–Time Relationship in Electroosmotic Flow. *J. Colloid Interface Sci.* **2003**, *261* (2), 402–410. https://doi.org/10.1016/S0021-9797(03)00142-5.

(38)   Sharma, P. K.; Rao, K. H. Adhesion of Paenibacillus Polymyxa on Chalcopyrite and Pyrite: Surface Thermodynamics and Extended DLVO Theory. *Colloids Surf. B Biointerfaces* **2003**, *29* (1), 21–38. https://doi.org/10.1016/S0927-7765(02)00180-7.

(39)   Chen, G.; Tallarek, U. Effect of Intraparticle Porosity and Double Layer Overlap on Electrokinetic Mobility in Multiparticle Systems. *Langmuir* **2003**, *19* (26), 10901–10908. https://doi.org/10.1021/la0355141.

(40)   Grimes, B. A.; Meyers, J. J.; Liapis, A. I. Determination of the Intraparticle Electroosmotic Volumetric Flow-Rate, Velocity and Peclet Number in Capillary Electrochromatography from Pore Network Theory. *J. Chromatogr. A* **2000**, *890* (1), 61–72.

(41)   Maier, R. S.; Nybo, E.; Seymour, J. D.; Codd, S. L. Electroosmotic Flow and Dispersion in Open and Closed Porous Media. *Transp. Porous Media* **2016**, *113* (1), 67–89. https://doi.org/10.1007/s11242-016-0680-4.

(42)   Tallarek, U.; Rapp, E.; Seidel-Morgenstern, A.; Van As, H. Electroosmotic Flow Phenomena in Packed Capillaries: From the Interstitial Velocities to Intraparticle and Boundary Layer Mass Transfer. *J. Phys. Chem. B* **2002**, *106* (49), 12709–12721. https://doi.org/10.1021/jp020605c.

(43)   Tallarek, U.; Rapp, E.; Van As, H.; Bayer, E. Electrokinetics in Fixed Beds: Experimental Demonstration of Electroosmotic Perfusion. *Angew. Chem. Int. Ed.* **2001**, *40* (9), 1684–1687. https://doi.org/10.1002/1521-3773(20010504)40:9<1684::AID-ANIE16840>3.0.CO;2-C.

(44)   Shan, Y.; Harms, H.; Wick, L. Y. Electric Field Effects on Bacterial Deposition and Transport in Porous Media. *Environ. Sci. Technol.* **2018**, *52* (24), 14294–14301. https://doi.org/10.1021/acs.est.8b03648.

(45)   Shaplro, A. P.; Probstein, R. F. Removal of Contaminants from Saturated Clay by Electroosmosis. *Environ. Sci. Technol.* **1993**, *27* (2), 283–291. https://doi.org/10.1021/es00039a007.

(46)   Pham, T. D.; Shrestha, R. A.; Sillanpää, M. Removal of Hexachlorobenzene and Phenanthrene from Clayey Soil by Surfactant-and Ultrasound-Assisted Electrokinetics. *J. Environ. Eng.* **2009**, *136* (7), 739–742. https://doi.org/10.1061/(ASCE)EE.1943-7870.0000203.

(47)   Jeon, C.-S.; Yang, J.-S.; Kim, K.-J.; Baek, K. Electrokinetic Removal of Petroleum Hydrocarbon from Residual Clayey Soil Following a Washing Process. *CLEAN - Soil Air Water* **2010**, *38* (2), 189–193. https://doi.org/10.1002/clen.200900190.

(48)   Kim, J.-H.; Han, S.-J.; Kim, S.-S.; Yang, J.-W. Effect of Soil Chemical Properties on the Remediation of Phenanthrene-Contaminated Soil by Electrokinetic-Fenton Process. *Chemosphere* **2006**, *63* (10), 1667–1676. https://doi.org/10.1016/j.chemosphere.2005.10.008.

(49)   Li, A.; Cheung, K. A.; Reddy, K. R. Cosolvent-Enhanced Electrokinetic Remediation of Soils Contaminated with Phenanthrene. *J. Environ. Eng.* **2000**, *126* (6), 527–533. https://doi.org/10.1061/(ASCE)0733-9372(2000)126:6(527).

(50)   Reddy, K. R. Technical Challenges to In-Situ Remediation of Polluted Sites. *Geotech. Geol. Eng.* **2010**, *28* (3), 211–221. https://doi.org/10.1007/s10706-008-9235-y.

(51)   Saichek, R. E.; Reddy, K. R. Effect of PH Control at the Anode for the Electrokinetic Removal of Phenanthrene from Kaolin Soil. *Chemosphere* **2003**, *51* (4), 273–287. https://doi.org/10.1016/S0045-6535(02)00849-4.

(52)   Rice, C. L.; Whitehead, R. Electrokinetic Flow in a Narrow Cylindrical Capillary. *J. Phys. Chem.* **1965**, *69* (11), 4017–4024. https://doi.org/10.1021/j100895a062.

(53)   Reddy, K. R.; Saichek, R. E. Enhanced Electrokinetic Removal of Phenanthrene from Clay Soil by Periodic Electric Potential Application. *J. Environ. Sci. Health Part A* **2004**, *39* (5), 1189–1212. https://doi.org/10.1081/ESE-120030326.

(54) Reddy, K. R.; Maturi, K.; Cameselle, C. Sequential Electrokinetic Remediation of Mixed Contaminants in Low Permeability Soils. *J. Environ. Eng.* **2009**, *135* (10), 989–998. https://doi.org/10.1061/(ASCE)EE.1943-7870.0000077.

(55) Kim, J. H.; Kim, S. S.; Yang, J. W. Role of Stabilizers for Treatment of Clayey Soil Contaminated with Phenanthrene through Electrokinetic-Fenton Process—Some Experimental Evidences. *Electrochimica Acta* **2007**, *53* (4), 1663–1670. https://doi.org/10.1016/j.electacta.2007.06.082.

(56) Hassan, I.; Mohamedelhassan, E.; Yanful, E. K.; Yuan, Z.-C. Sorption of Phenanthrene by Kaolin and Efficacy of Hydraulic versus Electroosmotic Flow to Stimulate Desorption. *J. Environ. Chem. Eng.* **2015**, *3* (4), 2301–2310. https://doi.org/10.1016/j.jece.2015.08.011.

(57) Vallano, P. T.; Remcho, V. T. Modeling Interparticle and Intraparticle (Perfusive) Electroosmotic Flow in Capillary Electrochromatography. *Anal. Chem.* **2000**, *72* (18), 4255–4265. https://doi.org/10.1021/ac0005969.

(58) Uskoković, V. Dynamic Light Scattering Based Microelectrophoresis: Main Prospects and Limitations. *J. Dispers. Sci. Technol.* **2012**, *33* (12), 1762–1786. https://doi.org/10.1080/01932691.2011.625523.

(59) Gavin, L.; Gillian, D. L. *Microbial Biofilms: Current Research and Applications*; Caister Academic: Norfolk, U.K., 2012.

(60) Martin, R. E.; Bouwer, E. J.; Hanna, L. M. Application of Clean-Bed Filtration Theory to Bacterial Deposition in Porous Media. *Environ. Sci. Technol.* **1992**, *26* (5), 1053–1058. https://doi.org/10.1021/es00029a028.

(61) Velasco-Casal, P.; Wick, L. Y.; Ortega-Calvo, J.-J. Chemoeffectors Decrease the Deposition of Chemotactic Bacteria during Transport in Porous Media. *Environ. Sci. Technol.* **2008**, *42* (4), 1131–1137. https://doi.org/10.1021/es071707p.

(62) Torkzaban, S.; Bradford, S. A.; Walker, S. L. Resolving the Coupled Effects of Hydrodynamics and DLVO Forces on Colloid Attachment in Porous Media. *Langmuir* **2007**, *23* (19), 9652–9660. https://doi.org/10.1021/la700995e.

(63) Kirby, B. J.; Hasselbrink, E. F. Zeta Potential of Microfluidic Substrates: 1. Theory, Experimental Techniques, and Effects on Separations. *ELECTROPHORESIS* **2004**, *25* (2), 187–202. https://doi.org/10.1002/elps.200305754.

(64) Bedrikovetsky, P.; Siqueira, F. D.; Furtado, C. A.; Souza, A. L. S. Modified Particle Detachment Model for Colloidal Transport in Porous Media. *Transp. Porous Media* **2011**, *86* (2), 353–383. https://doi.org/10.1007/s11242-010-9626-4.

(65) Van Oss, C. J.; Good, R. J.; Chaudhury, M. K. The Role of van Der Waals Forces and Hydrogen Bonds in "Hydrophobic Interactions" between Biopolymers and Low Energy Surfaces. *J. Colloid Interface Sci.* **1986**, *111* (2), 378–390. https://doi.org/10.1016/0021-9797(86)90041-X.

(66) Vilinska, A.; Rao, K. H. Surface Thermodynamics and Extended DLVO Theory of Leptospirillum Ferrooxidans Cells' Adhesion on Sulfide Minerals. *Min. Metall. Explor.* **2011**, *28* (3), 151–158. https://doi.org/10.1007/BF03402248.

(67) Fowkes, F. M. Attractice Forces at Interfaces. *Ind. Eng. Chem.* **1964**, *56* (12), 40–52. https://doi.org/10.1021/ie50660a008.

(68) Brown, D. G.; Jaffé, P. R. Effects of Nonionic Surfactants on the Cell Surface Hydrophobicity and Apparent Hamaker Constant of a *Sphingomonas* Sp. *Environ. Sci. Technol.* **2006**, *40* (1), 195–201. https://doi.org/10.1021/es051183y.

(69) Van Oss, C. J.; Chaudhury, M. K.; Good, R. J. Interfacial Lifshitz-van Der Waals and Polar Interactions in Macroscopic Systems. *Chem. Rev.* **1988**, *88* (6), 927–941.

(70) Meylheuc, T.; Methivier, C.; Renault, M.; Herry, J.-M.; Pradier, C.-M.; Bellon-Fontaine, M. N. Adsorption on Stainless Steel Surfaces of Biosurfactants Produced by Gram-Negative and Gram-Positive Bacteria: Consequence on the Bioadhesive Behavior of Listeria Monocytogenes. *Colloids Surf. B Biointerfaces* **2006**, *52* (2), 128–137. https://doi.org/10.1016/j.colsurfb.2006.04.016.

(71) Ward, M. D.; Buttry, D. A. In Situ Interfacial Mass Detection with Piezoelectric Transducers. *Science* **1990**, *249* (4972), 1000–1007.

(72)   Reviakine, I.; Johannsmann, D.; Richter, R. P. Hearing What You Cannot See and Visualizing What You Hear: Interpreting Quartz Crystal Microbalance Data from Solvated Interfaces. *Anal. Chem.* **2011**, *83* (23), 8838–8848. https://doi.org/10.1021/ac201778h.

(73)   Sauerbrey, G. Verwendung von Schwingquarzen zur Wägung dünner Schichten und zur Mikrowägung. *Z. Für Phys.* **1959**, *155* (2), 206–222. https://doi.org/10.1007/BF01337937.

(74)   Gutman, J.; Walker, S. L.; Freger, V.; Herzberg, M. Bacterial Attachment and Viscoelasticity: Physicochemical and Motility Effects Analyzed Using Quartz Crystal Microbalance with Dissipation (QCM-D). *Environ. Sci. Technol.* **2013**, *47* (1), 398–404. https://doi.org/10.1021/es303394w.

(75)   Marcus, I. M.; Herzberg, M.; Walker, S. L.; Freger, V. Pseudomonas Aeruginosa Attachment on QCM-D Sensors: The Role of Cell and Surface Hydrophobicities. *Langmuir* **2012**, *28* (15), 6396–6402. https://doi.org/10.1021/la300333c.

(76)   Kao, W.-L.; Chang, H.-Y.; Lin, K.-Y.; Lee, Y.-W.; Shyue, J.-J. Effect of Surface Potential on the Adhesion Behavior of NIH3T3 Cells Revealed by Quartz Crystal Microbalance with Dissipation Monitoring (QCM-D). *J. Phys. Chem. C* **2017**, *121* (1), 533–541. https://doi.org/10.1021/acs.jpcc.6b11217.

(77)   Sharma, P. K.; Hanumantha Rao, K. Adhesion of Paenibacillus Polymyxa on Chalcopyrite and Pyrite: Surface Thermodynamics and Extended DLVO Theory. *Colloids Surf. B Biointerfaces* **2003**, *29* (1), 21–38. https://doi.org/10.1016/S0927-7765(02)00180-7.

(78)   Van Oss, C. J.; Giese, R. F.; Costanzo, P. M. DLVO and Non-DLVO Interactions in Hectorite. *Clays Clay Min.* **1990**, *38* (2), 151–159.

(79)   Redman, J. A.; Walker, S. L.; Elimelech, M. Bacterial Adhesion and Transport in Porous Media: Role of the Secondary Energy Minimum. *Environ. Sci. Technol.* **2004**, *38* (6), 1777–1785. https://doi.org/10.1021/es0348871.

(80)   Goldman, A. J.; Cox, R. G.; Brenner, H. Slow Viscous Motion of a Sphere Parallel to a Plane Wall—II Couette Flow. *Chem. Eng. Sci.* **1967**, *22* (4), 653–660. https://doi.org/10.1016/0009-2509(67)80048-4.

(81)   Solomentsev, Y.; Böhmer, M.; Anderson, J. L. Particle Clustering and Pattern Formation during Electrophoretic Deposition: A Hydrodynamic Model. *Langmuir* **1997**, *13* (23), 6058–6068.

(82)   Probstein, R. F. *Physicochemical Hydrodynamics: An Introduction*; John Wiley & Sons, 2005.

(83)   Mastral, A. M.; Callén, M. S. A Review on Polycyclic Aromatic Hydrocarbon (PAH) Emissions from Energy Generation. *Environ. Sci. Technol.* **2000**, *34* (15), 3051–3057. https://doi.org/10.1021/es001028d.

(84)   Parajulee, A.; Wania, F. Evaluating Officially Reported Polycyclic Aromatic Hydrocarbon Emissions in the Athabasca Oil Sands Region with a Multimedia Fate Model. *Proc. Natl. Acad. Sci.* **2014**, *111* (9), 3344–3349. https://doi.org/10.1073/pnas.1319780111.

(85)   Peachey, R. B. J. Tributyltin and Polycyclic Aromatic Hydrocarbon Levels in Mobile Bay, Alabama: A Review. *Mar. Pollut. Bull.* **2003**, *46* (11), 1365–1371. https://doi.org/10.1016/S0025-326X(03)00373-4.

(86)   Molina, M. C.; González, N.; Bautista, L. F.; Sanz, R.; Simarro, R.; Sánchez, I.; Sanz, J. L. Isolation and Genetic Identification of PAH Degrading Bacteria from a Microbial Consortium. *Biodegradation* **2009**, *20* (6), 789–800. https://doi.org/10.1007/s10532-009-9267-x.

(87)   Chen, J.; Wong, M. H.; Wong, Y. S.; Tam, N. F. Y. Multi-Factors on Biodegradation Kinetics of Polycyclic Aromatic Hydrocarbons (PAHs) by Sphingomonas Sp. a Bacterial Strain Isolated from Mangrove Sediment. *Mar. Pollut. Bull.* **2008**, *57* (6–12), 695–702. https://doi.org/10.1016/j.marpolbul.2008.03.013.

(88)   Harms, H.; Wick, L. Y. Mobilisation of Organic Compounds and Iron by Microorganisms. In *Physicochemical Kinetics and Transport at Biointerfaces*; Leeuwen, Herman. P. van, Köster, W., Eds.; John Wiley & Sons, Ltd: Chichester, UK, 2004; pp 401–444. https://doi.org/10.1002/0470094044.ch9.

(89) Morelis, S.; van Noort, P. C. M. Kinetics of Phenanthrene Desorption from Activated Carbons to Water. *Chemosphere* **2008**, *71* (11), 2044–2049. https://doi.org/10.1016/j.chemosphere.2008.01.048.

(90) Ahn, S.; Werner, D.; Karapanagioti, H. K.; McGlothlin, D. R.; Zare, R. N.; Luthy, R. G. Phenanthrene and Pyrene Sorption and Intraparticle Diffusion in Polyoxymethylene, Coke, and Activated Carbon. *Environ. Sci. Technol.* **2005**, *39* (17), 6516–6526. https://doi.org/10.1021/es050113o.

(91) Ho, Y. S.; Ng, J. C. Y.; McKay, G. Kinetics of Pollutant Sorption by Biosorbents. *Sep. Purif. Methods* **2000**, *29* (2), 189–232. https://doi.org/10.1081/SPM-100100009.

(92) Huang, W.; Weber, W. J. Thermodynamic Considerations in the Sorption of Organic Contaminants by Soils and Sediments. 1. The Isosteric Heat Approach and Its Application to Model Inorganic Sorbents. *Environ. Sci. Technol.* **1997**, *31* (11), 3238–3243. https://doi.org/10.1021/es970230m.

(93) Doke, K. M.; Khan, E. M. Adsorption Thermodynamics to Clean up Wastewater; Critical Review. *Rev. Environ. Sci. Biotechnol.* **2013**, *12* (1), 25–44. https://doi.org/10.1007/s11157-012-9273-z.

(94) Shan, Y.; Qin, J.; Harms, H.; Wick, L. Y. Electrokinetic Effects on the Interaction of Phenanthrene with Geo-Sorbents. *Chemosphere* **2019**, 125161.

(95) Do, D. D. *Adsorption Analysis: Equilibria and Kinetics*; Series on chemical engineering; Imperial College Press: London, 1998.

(96) *Adsorption: Theory, Modeling, and Analysis*; Tóth, J., Ed.; Surfactant science series; Marcel Dekker: New York, 2002.

(97) Tran, H. N.; You, S.-J.; Hosseini-Bandegharaei, A.; Chao, H.-P. Mistakes and Inconsistencies Regarding Adsorption of Contaminants from Aqueous Solutions: A Critical Review. *Water Res.* **2017**, *120*, 88–116. https://doi.org/10.1016/j.watres.2017.04.014.

(98) Ghosal, P. S.; Gupta, A. K. An Insight into Thermodynamics of Adsorptive Removal of Fluoride by Calcined Ca–Al–(NO3) Layered Double Hydroxide. *RSC Adv.* **2015**, *5* (128), 105889–105900. https://doi.org/10.1039/C5RA20538G.

(99) Kopinke, F.-D.; Georgi, A.; Goss, K.-U. Comment on "Mistakes and Inconsistencies Regarding Adsorption of Contaminants from Aqueous Solution: A Critical Review, Published by Tran et al. [Water Research 120, 2017, 88–116]." *Water Res.* **2018**, *129*, 520–521. https://doi.org/10.1016/j.watres.2017.09.055.

(100) Tanner, F. W. *Shape and Size of Bacterial Cells: Bacteriology*; John Wiley and Sons, Inc., New York, 1948.

(101) Harms, H.; Wick, L. Y. Dispersing Pollutant-Degrading Bacteria in Contaminated Soil without Touching It. *Eng. Life Sci.* **2006**, *6* (3), 252–260. https://doi.org/10.1002/elsc.200620122.

(102) Semple, K. T.; Doick, K. J.; Wick, L. Y.; Harms, H. Microbial Interactions with Organic Contaminants in Soil: Definitions, Processes and Measurement. *Environ. Pollut.* **2007**, *150* (1), 166–176. https://doi.org/10.1016/j.envpol.2007.07.023.

(103) Sinton, D.; Li, D. Electroosmotic Velocity Profiles in Microchannels. *Colloids Surf. Physicochem. Eng. Asp.* **2003**, *222* (1–3), 273–283.

(104) Ai, L.; Jiang, J. Removal of Methylene Blue from Aqueous Solution with Self-Assembled Cylindrical Graphene–Carbon Nanotube Hybrid. *Chem. Eng. J.* **2012**, *192*, 156–163. https://doi.org/10.1016/j.cej.2012.03.056.

(105) Liu, Y. Is the Free Energy Change of Adsorption Correctly Calculated? *J. Chem. Eng. Data* **2009**, *54* (7), 1981–1985. https://doi.org/10.1021/je800661q.

(106) Drost-Hansen, W. Water at Biological Interfaces - Structural and Functional Aspects. *Phys. Chem. Liq.* **1978**, *7* (3–4), 243–348. https://doi.org/10.1080/00319107808084734.

(107) Mader, B. T.; Uwe-Goss, K.; Eisenreich, S. J. Sorption of Nonionic, Hydrophobic Organic Chemicals to Mineral Surfaces. *Environ. Sci. Technol.* **1997**, *31* (4), 1079–1086. https://doi.org/10.1021/es960606g.

(108) Soni, K. A.; Balasubramanian, A. K.; Beskok, A.; Pillai, S. D. Zeta Potential of Selected Bacteria in Drinking Water When Dead, Starved, or Exposed to Minimal and Rich Culture Media. *Curr. Microbiol.* **2008**, *56* (1), 93–97. https://doi.org/10.1007/s00284-007-9046-z.

(109) Van Loosdrecht, M. C.; Lyklema, J.; Norde, W.; Schraa, G.; Zehnder, A. J. Electrophoretic Mobility and Hydrophobicity as a Measured to Predict the Initial Steps of Bacterial Adhesion. *Appl. Environ. Microbiol.* **1987**, *53* (8), 1898–1901.

(110) Stenström, T. A. Bacterial Hydrophobicity, an Overall Parameter for the Measurement of Adhesion Potential to Soil Particles. *Appl. Environ. Microbiol.* **1989**, *55* (1), 142–147.

(111) Vane, L. M.; Zang, G. M. Effect of Aqueous Phase Properties on Clay Particle Zeta Potential and Electro-Osmotic Permeability: Implications for Electro-Kinetic Soil Remediation Processes. *J. Hazard. Mater.* **1997**, *55* (1–3), 1–22.

(112) Yukselen, Y.; Kaya, A. Zeta Potential of Kaolinite in the Presence of Alkali, Alkaline Earth and Hydrolyzable Metal Ions. *Water. Air. Soil Pollut.* **2003**, *145* (1–4), 155–168.

(113) Simoni, S. F.; Bosma, T. N. P.; Harms, H.; Zehnder, A. J. B. Bivalent Cations Increase Both the Subpopulation of Adhering Bacteria and Their Adhesion Efficiency in Sand Columns. *Environ. Sci. Technol.* **2000**, *34* (6), 1011–1017. https://doi.org/10.1021/es990476m.

(114) Simoni, S. F.; Harms, H.; Bosma, T. N. P.; Zehnder, A. J. B. Population Heterogeneity Affects Transport of Bacteria through Sand Columns at Low Flow Rates. *Environ. Sci. Technol.* **1998**, *32* (14), 2100–2105. https://doi.org/10.1021/es970936g.

8. Appendix

8.1 Declaration of Independent Work

Declaration of independent work

I hereby declare that

- I have written this thesis autonomously incorporating my own ideas and judgments; no other resources were used without being stated. The quotations from other work have been marked accordingly in the text and full reference of the source has been provided in the proper way.

- All persons are listed who supported me for the selection and evaluation of the materials for my thesis; the contributions of co-authors are listed in "Author contributions of published articles" (5.2). No other persons have provided support and thereby contributed to the thesis, and no third party has received direct or indirect financial benefits in goods and services for work that stands in relation to the work presented in the thesis.

- This thesis has not been submitted in an equal or similar form for examination for the degree of doctorate or any other degree at another academic institution, and has not been published.

Place   Date   Signature